# Mutagenic Effects of Environmental Contaminants

# ENVIRONMENTAL SCIENCES

## An Interdisciplinary Monograph Series

### EDITORS



ARTHUR C. STERN, editor, AIR POLLUTION, Second Edition. In three volumes, 1968

L. FISHBEIN, W. G. FLAMM, and H. L. FALK, CHEMICAL MUTAGENS: Environmental Effects on Biological Systems, 1970

DOUGLAS H. K. LEE and DAVID MINARD, editors, PHYSIOLOGY, ENVIRONMENT, AND MAN, 1970

KARL D. KRYTER, THE EFFECTS OF NOISE ON MAN, 1970

R. E. MUNN, BIOMETEOROLOGICAL METHODS, 1970

M. M. KEY, L. E. KERR, and M. BUNDY, PULMONARY REACTIONS TO COAL DUST: "A Review of U. S. Experience," 1971

DOUGLAS H. K. LEE, editor, METALLIC CONTAMINANTS AND HUMAN HEALTH, 1972

DOUGLAS H. K. LEE, editor, ENVIRONMENTAL FACTORS IN RESPIRATORY DISEASE, 1972

H. ELDON SUTTON and MAUREEN I. HARRIS, editors, MUTAGENIC EFFECTS OF ENVIRONMENTAL CONTAMINANTS, 1972

*In preparation*

MOHAMED K. YOUSEF, STEVEN M. HORVATH, and ROBERT W. BULLARD, PHYSIOLOGICAL ADAPTATIONS: Desert and Mountain

DOUGLAS H. K. LEE and PAUL KOTIN, editors, MULTIPLE FACTORS IN THE CAUSATION OF ENVIRONMENTALLY INDUCED DISEASE, 1972

*Fogarty International Center Proceedings No. 10*

# Mutagenic Effects of Environmental Contaminants

***Scientific Editors***

**H. Eldon Sutton**

DEPARTMENT OF ZOOLOGY
UNIVERSITY OF TEXAS AT AUSTIN
AUSTIN, TEXAS

**Maureen I. Harris**

THE JOHN E. FOGARTY INTERNATIONAL CENTER
NATIONAL INSTITUTES OF HEALTH
BETHESDA, MARYLAND

Sponsored by
The John E. Fogarty International Center
for Advanced Study in the Health Sciences
National Institutes of Health
Bethesda, Maryland
and
The National Institute of Environmental Health Sciences
National Institutes of Health
Research Triangle Park, North Carolina

Academic Press
New York and London 1972

ACADEMIC PRESS, INC.
111 Fifth Avenue, New York, New York 10003

*United Kingdom Edition published by*
ACADEMIC PRESS, INC. (LONDON) LTD.
24/28 Oval Road,
London NW1 7DD

LIBRARY OF CONGRESS CATALOG CARD NUMBER: 77-189673



THE OPINIONS EXPRESSED IN THIS BOOK ARE THOSE OF THE RESPECTIVE AUTHORS AND DO NOT NECESSARILY REPRESENT THOSE OF THE DEPARTMENT OF HEALTH, EDUCATION, AND WELFARE.

PRINTED IN THE UNITED STATES OF AMERICA

# CONTENTS

ORGANIZING PANEL . . . . . . . . . . . . . . . . . . . . . . . ix
CONTRIBUTORS . . . . . . . . . . . . . . . . . . . . . . . xi
PREFACE . . . . . . . . . . . . . . . . . . . . . . . xiii

**INTRODUCTION: GENETIC TOXICOLOGY**
*John W. Drake*
Text . . . . . . . . . . . . . . . . . . . . . . . 1

**GENE MUTATION AS A CAUSE OF HUMAN DISEASE** . . . . . . .
*Barton Childs, Stuart M. Miller, and Alexander G. Bearn*
Text . . . . . . . . . . . . . . . . . . . . . . . 3
References . . . . . . . . . . . . . . . . . . . . . 14

**THE MOLECULAR BASIS OF MUTATION**
*John W. Drake and W. Gary Flamm*
The Genetic Material . . . . . . . . . . . . . . . . . 16
Replication of Genetic Material . . . . . . . . . . . . 16
Utilization of Genetic Information . . . . . . . . . . . 17
The Nature of Mutations . . . . . . . . . . . . . . . 19
Mechanisms of Mutagenesis . . . . . . . . . . . . . . 21
The Effects of Mutation on the Organism . . . . . . . . 24

**MONITORING OF CHEMICAL MUTAGENS IN OUR ENVIRONMENT**
*Heinrich V. Malling*
Introduction . . . . . . . . . . . . . . . . . . . . 27
Problems in Testing for Mutagenic Activity . . . . . . . . 30
1. Comparative Mutagenesis . . . . . . . . . . . . 32
2. Evaluation of Results from Mutagenesis Testing . . . . . 33
3. Weak Mutagenic Activity: A Major Problem Area . . . . 36
Conclusion . . . . . . . . . . . . . . . . . . . . . 37
References . . . . . . . . . . . . . . . . . . . . . 37

**THE DETECTION OF MUTATIONS WITH NON-MAMMALIAN SYSTEMS**
*Frederick J. de Serres*
Prokaryote Systems . . . . . . . . . . . . . . . . . 41
1. DNA Transformation . . . . . . . . . . . . . 41
2. Bacteriophage . . . . . . . . . . . . . . . . 42
3. Bacteria . . . . . . . . . . . . . . . . . . 43

Eukaryote Systems . . . . . . . . . . . . . . . . . . . 44
1. Ascomycetes . . . . . . . . . . . . . . . . . . 44
2. Drosophila . . . . . . . . . . . . . . . . . . 51
3. Habrobracon . . . . . . . . . . . . . . . . . . 52
4. Tradescantia . . . . . . . . . . . . . . . . . . 53
References . . . . . . . . . . . . . . . . . . . . . 54

**A BACTERIAL SYSTEM FOR DETECTING MUTAGENS AND CARCINOGENS**
*Bruce N. Ames*
Advantages of Using a Particular Set of Four Strains We Have Developed in *Salmonella typhimurium* . . . . . . . . . . . 58
Frameshift Mutagens and Their Relation to Polycyclic Hydrocarbons . . . . . . . . . . . . . . . . . . . . 61
Agents That Have Been Shown to Be Mutagenic Using These Strains . . . . . . . . . . . . . . . . . . 62
Reasons Why a Compound That Is Mutagenic or Carcinogenic in Humans May Be Missed in a Bacterial Screen . . . . . . . 63
Reasons That a Compound That Is Mutagenic for Bacteria May Not Be Mutagenic for Humans . . . . . . . . . . . . . 64
References . . . . . . . . . . . . . . . . . . . . . . 66

**THE NEED TO DETECT CHEMICALLY INDUCED MUTATIONS IN EXPERIMENTAL ANIMALS**
*Marvin S. Legator*
The Need to Use Mammalian Systems for Mutagenicity Testing . . 69
Animal (Mammalian) Test Systems . . . . . . . . . . . . 70
Impact of Mutagenicity Screening on Carcinogenicity Studies. . . 74
Insensitivity of Animal Data . . . . . . . . . . . . . . 75
Definitive Animal Experiments – The Only Practical Means of Eliminating Mutagenic Agents . . . . . . . . . . 76
References . . . . . . . . . . . . . . . . . . . . . . 79

**CHROMOSOME MUTATIONS IN MAN**
*Margery W. Shaw*
Introduction . . . . . . . . . . . . . . . . . . . . . 81
Human Chromosomal Polymorphism . . . . . . . . . . . 82
Characterizing Human Chromosomal Mutations . . . . . . . . 82
Monitoring Human Chromosomal Mutations . . . . . . . . . 83
Detecting an Increase in Human Chromosomal Mutations . . . . 95
References . . . . . . . . . . . . . . . . . . . . . . 97

**THE DETECTION OF INCREASED MUTATION RATES IN HUMAN POPULATIONS**
*James V. Neel*
Introduction . . . . . . . . . . . . . . . . . . . . . 99

The Approaches to Monitoring Human Populations . . . . . . . 100
1. The Use of Population Characteristics . . . . . . . . . 101
2. The Use of Sentinel Phenotypes . . . . . . . . . . . . 101
3. The Use of Biochemical and/or Chromosomal Mutations . 105
Some Problems in Monitoring for Mutations
at the Chemical Level . . . . . . . . . . . . . . . . . . 107
1. Primarily Technical Questions . . . . . . . . . . . . 107
2. Primarily Theoretical Issues . . . . . . . . . . . . . 114
Concluding Remarks . . . . . . . . . . . . . . . . . . . 116
References . . . . . . . . . . . . . . . . . . . . . . . 118

**MONITORING SOMATIC MUTATIONS IN HUMAN POPULATIONS**
*H. Eldon Sutton*
The Frequency of Point Mutations . . . . . . . . . . . . . 122
Somatic Point Mutations . . . . . . . . . . . . . . . . . 123
Repair of Somatic Mutations . . . . . . . . . . . . . . . 125
Sentinel Phenotypes for Somatic Mutation . . . . . . . . . 126
Conclusion . . . . . . . . . . . . . . . . . . . . . . . 127
References . . . . . . . . . . . . . . . . . . . . . . . 128

**PESTICIDAL, INDUSTRIAL, FOOD ADDITIVE, AND DRUG MUTAGENS**
*Lawrence Fishbein*
Introduction . . . . . . . . . . . . . . . . . . . . . . 129
I. Pesticidal Mutagens . . . . . . . . . . . . . . . . . 132
II. Industrial Mutagens . . . . . . . . . . . . . . . . . 142
III. Food and Feed Additive and Naturally Occurring
Mutagens . . . . . . . . . . . . . . . . . . . . . 153
IV. Drug Mutagens . . . . . . . . . . . . . . . . . . . 160
References . . . . . . . . . . . . . . . . . . . . . . . 170

**MUTAGENICITY OF BIOLOGICALS**
*Warren W. Nichols*
Text . . . . . . . . . . . . . . . . . . . . . . . . . . 171
Conclusion . . . . . . . . . . . . . . . . . . . . . . . 175
References . . . . . . . . . . . . . . . . . . . . . . . 176

**POSSIBLE RELATIONSHIPS BETWEEN MUTAGENESIS AND CARCINOGENESIS**
*Isaac Berenblum*
Text . . . . . . . . . . . . . . . . . . . . . . . . . . 177
References . . . . . . . . . . . . . . . . . . . . . . . 182

**INTERRELATIONS BETWEEN CARCINOGENICITY, MUTAGENICITY, AND TERATOGENICITY**
*James G. Wilson*
Text . . . . . . . . . . . . . . . . . . . . . . . 185
Conclusion . . . . . . . . . . . . . . . . . . . . 194
References . . . . . . . . . . . . . . . . . . . . 194

# ORGANIZING PANEL

*Peter G. Condliffe,* Chief, Conference and Seminar Program Branch, Fogarty International Center for Advanced Study in the Health Sciences, National Institutes of Health, Bethesda, Maryland

*W. Gary Flamm*, Cell Biology Branch, National Institute of Environmental Health Sciences, National Institutes of Health, Research Triangle Park, North Carolina

*Douglas H. K. Lee*, Associate Director, National Institute of Environmental Health Sciences, National Institutes of Health, Research Triangle Park, North Carolina

*Marvin S. Legator*, Chief, Cell Biology Branch, Division of Nutrition, Food and Drug Administration, Washington, D. C.

*H. Eldon Sutton*, Chairman, Department of Zoology, University of Texas at Austin, Austin, Texas

# CONTRIBUTORS

*Bruce N. Ames*, Department of Biochemistry, University of California, Berkeley, California

*Alexander G. Bearn*, Professor of Medicine, Cornell University Medical College, New York, New York

*Isaac Berenblum*,* Scholar-in-Residence, Fogarty International Center, National Institutes of Health, Bethesda, Maryland

*Barton Childs*, Department of Pediatrics, The Johns Hopkins Hospital, Baltimore, Maryland

*Frederick J. de Serres*, Biology Division, Oak Ridge National Laboratory, Oak Ridge, Tennessee

*John W. Drake*, Department of Microbiology, University of Illinois, Urbana, Illinois

*Lawrence Fishbein*, Chief, Analytical and Scientific Chemistry Branch, National Institute of Environmental Health Sciences, Research Triangle Park, North Carolina

*W. Gary Flamm*, Cell Biology Branch, National Institute of Environmental Health Sciences, Research Triangle Park, North Carolina

*Marvin S. Legator*, Chief, Cell Biology Branch, Division of Nutrition, Food and Drug Administration, Washington, D. C.

*Heinrich V. Malling*, Biology Division, Oak Ridge National Laboratory, Oak Ridge, Tennessee

*Stuart M. Miller*, Fellow, The New York Hospital, New York, New York

*James V. Neel*, Department of Human Genetics, University of Michigan Medical School, Ann Arbor, Michigan

*Warren W. Nichols*, Department of Cytogenetics, Institute for Medical Research, Camden, New Jersey

*Margery W. Shaw*, Section of Medical Genetics, M. D. Anderson Hospital and Tumor Institute, The University of Texas Medical Center, Houston, Texas

*Current address – The Weizmann Institute, Rehovot, Israel

*H. Eldon Sutton*, Chairman, Department of Zoology, University of Texas at Austin, Austin, Texas

*James G. Wilson*, The Children's Hospital Research Foundation, University of Cincinnati College of Medicine, Cincinnati, Ohio

# PREFACE

All persons live in a global ecosystem consisting of life cycles that tie man to microorganisms, plants, and animals – a very immediate relationship exists between man and his environment. While man can change his environment almost at will, he must be able to weigh knowledgeably the costs of change against the benefits. Much of the prosperity enjoyed today is due to technology, and society at large is now very dependent on it. However, many experts feel that the synthetic materials, chemical compounds, and drugs that are being marketed are becoming, if only because of their increased numbers, more deleterious in their genetic effects on living organisms. The 1969-1970 HEW Task Force on Research Planning in Environmental Health Science noted that chemicals which are known or suspected of being harmful to human beings are being produced and marketed at an increasing rate. New products are being introduced into our society faster than their safety can be ascertained. Although these substances may aid man in developing a better life, they may also have insidious effects on his health. Certain people are not selected out–the whole community of man can be affected, even generations yet unborn–and the perpetuation of genetic effects to future generations emphasizes the importance of the problem. Detrimental mutations can be easily induced in the laboratory in a variety of biological systems, attesting to the occurrence of induced genetic effects in all life forms.

Both governmental agencies and public groups have expressed concern about the deleterious effects of environmental contaminants, and their interest has stimulated the establishment of scientific and organizational systems to investigate and control man's manipulation of the environment. A variety of scientists are being called upon by these groups to advise in these new areas which cross older scientific disciplines. In addition, administrators of environmental programs are expected to have some scientific expertise in order to effectively manage their programs. The announcement of a United Nations Conference on the Human Environment to be held in Stockholm in June, 1972, indicates the worldwide concern for this problem.

To try to meet the needs of these groups, the Fogarty International Center for Advanced Study in the Health Sciences and the National Institute of Environmental Health Sciences, both of the National Institutes of Health, decided to prepare a comprehensive reference book on environmental contaminants and their genetic effects.* A panel of experts brought working drafts to a three-day meeting

*Three other books have been sponsored by the Fogarty International Center and the National Institute of Environmental Health Sciences: "Environmental Factors in Respiratory Disease," "Metallic Contaminants and Human Health," and "Multiple Factors in the Causation of Environmentally Induced Disease." They will also appear in the Environmental Sciences Series published by Academic Press.

held at the Fogarty Center, where the drafts were discussed, amended, and integrated. The editing which followed attempted to simplify the text somewhat for the nonspecialist without compromising the quality of the scientific material for the expert. The resulting volume covers the mutagenic effects of environmental contaminants by examining the current burden of human genetic disease; the biochemical mechanisms of mutation; the practical and feasible tests, using a variety of organisms, for screening potential mutagenic agents; methods of determining the human mutation rate; a catalog of mutagens which are currently in use commercially; and the possible interrelationships between mutagenesis, carcinogenesis, and teratogenesis. We hope that this book will serve as a comprehensive reference for those interested in the current knowledge and techniques in this area and as a basis for recommendations for future tests and controls of mutagenic contaminants.

Maureen I. Harris
H. Eldon Sutton

# Mutagenic Effects of Environmental Contaminants

# INTRODUCTION: GENETIC TOXICOLOGY

*John W. Drake*

The modern heritage, whether one exaults or abhors it, will largely determine the human condition for centuries to come. This heritage comes to us in two forms: as learned elements, largely available in the form of written symbols or traditional skills; and as genetically inherited elements, available in the form of the diverse combinations which constitute the gene pool. The physical nature of the earth, particularly within those first few feet above and below its surface which comprise man's environment, has changed in the past hundred years or so at a rate without historical precedent. This change, largely engendered by man himself, presents both cultural and genetic hazards. Just how great is the potential for genetic damage has only become evident in very recent years, and usually only to professional geneticists. It is the purpose of this book to assess the nature of this potential, to review the scientific and organizational systems which are available now or else will be needed in the immediate future, and to offer, as best we can, a prognosis for protection.

The conceptual and technical problems are imposing, although certainly no more so than others which have been faced within this century alone. In relation to the knowledge required to evaluate man's genetic condition unequivocally, our present knowledge is tiny. We have little understanding of the normal function of most human genetic material, little knowledge of the overall human mutation rate, and even less knowledge of the optimal human mutation rate. We know rather little about which chemicals can penetrate to human germ cells, and not much more about what probably happens when they do penetrate. On the other hand, man is rapidly devising and widely disseminating many new and highly reactive chemicals, often without any prior knowledge whatsoever about their potential genetic toxicology. The shortsightedness of this activity becomes abundantly clear when one considers the already high price which man pays for his burden of mutations. Any increase in the human mutation rate will affect not only those children currently being conceived, but also their descendents down through time.

Devising reliable safeguards against genetic damage will require considerable advances on several fronts. Basic knowledge concerning the molecular nature of the mutation process and also concerning human genetics generally is obviously required in order to define the nature of the genetic target. Monitoring and testing systems must be developed and perfected in order to detect potential mutagenicity of suspect substances, and we must soon come to grips with the problem of how to utilize information gained from these test systems. A realistic rationale must be devised for genetic cost accounting, so that potential dangers can be compared with potential benefits. And finally, suitable parascientific organizations must be brought into existence on both the national and the international levels, to coordinate all of these efforts and to seek enforcement of scientific judgements within the wider political and social arena.

Even for those who have grappled with this problem for years, today is just a beginning.

# GENE MUTATION AS A CAUSE OF HUMAN DISEASE*

*Barton Childs, Stuart M. Miller, and Alexander G. Bearn*

Genes control the specificity, amount, and developmental characteristics of proteins, which in turn determine normal growth and development as well as normal body function. Mutations in human genes may be innocuous, contributing simply to normal variability, or they may be damaging, causing an alteration in the function of the character under gene control. Ideally, an assessment of the effect of mutation on human beings should include the effects of each mutation in each gene on the protein it specifies. A beginning of this kind of description has been made for more than 60 human genes (McKusick, 1970). The mutations result in changes in the physical and chemical properties of the proteins involved, and most appear to be harmless to the people possessing them. Some of the variants for some of the genes are quite common, suggesting that these mutations may have a selective advantage, while others are rare. For some of these genes, very large numbers of different mutants are known. For example, there are more than 100 known variants of hemoglobin whose amino acid substitutions are known. Twenty percent of these cause disease, but this probably represents bias in the choice of individuals in whom the search was made for an abnormal hemoglobin, suggesting that, if one were to sample the total population, the vast bulk of mutants would be harmless. At least 50 percent of human genes show common variation, and each individual is heterozygous for perhaps 15 percent of his own genes (Lewontin, 1967). Because of the limitations of the techniques so far used to detect this variability, it seems likely that these figures are underestimates.

---

*This work was aided by USPHS Grant #HD 00486.

Study of the physical properties of many proteins is not possible, and to detect variation one must depend upon other properties, for example enzyme activity. There are now more than 90 human variations, nearly all diseases, some seriously crippling, some lethal, which are characterized by marked diminution in the activity of enzymes (McKusick, 1970).

In most instances, there is no descriptive information about the enzyme except for its reduction in activity, but when direct observations of other properties of the proteins could be made, abnormal proteins have been found. Thus, the extent to which these disorders may represent mutations of regulatory elements akin to those studied in bacteria has not been established. There is a large range in the severity of the disorders resulting from these variations in enzyme activity. This suggests that there is a large undiscovered reservoir of variants at both ends of the spectrum, namely those which constitute normal variation as well as an unpredictable number whose effects are lethal.

There are more than 1600 human genetical variants known, and many are seen as morphological differences (McKusick, 1968). The latter are the result of distortions due to arrest of, or other interference with, intrauterine development. In contrast to the inborn errors of metabolism, most of them segregate as dominants and their biochemical attributes are unknown.

Some diseases tend to recur in families, but seem not to be due to simple inheritance of a mutant gene. Examples of these diseases are schizophrenia and diabetes. The familial distributions of these diseases suggest that they originate in the simultaneous action of several genes. Nothing is known about the nature of these mutations, the number or the function of such genes. The disorders due to mutant genes showing simple Mendelian inheritance tend to be individually rare, varying downward, in general, from frequencies of one per 10,000 persons, while multigenic disorders may be quite common. For example, diabetes occurs in about one in every 110 people in the United States.

Human genetic material is also susceptible to mutational events which cause changes in the number of chromosomes or alterations in their structure (Table I). These may occur in meiosis during the formation of eggs and sperm, after which the chromosomal abnormality is present in all of the cells of the organism, or in mitosis (simple cell division), after which the abnormality appears in only some of the cells. In the latter case, the individual is a genetic mosaic. Many of the meiotic chromosome abnormalities are lethal, while others are devastating to the health of their possessors. Approximately 20 percent of all conceptuses have chromosome abnormalities (Carr, 1969), but most of these die _in utero_, since only 0.5 percent of liveborn infants have chromosome abnormalities (Ratcliffe et al., 1970). If a mitotic aberration occurs very early in the development of the embryo, mosaicism is extensive and the infant may be damaged; if late in embryogenesis, mosaicism may be insignificant. Human cells tolerate autosomal deletions poorly, while the addition of as much as a single autosome, although seriously damaging, may permit life. Excesses or deletions of sex chromosomes, on the other hand, are relatively well tolerated. A few such individuals are effectively normal and may even be fertile, but careful assessment of the physical and behavioral

Table I. A Classification of Chromosome Abnormalities

| ERRORS OF NUMBER | ERRORS OF STRUCTURE |
|---|---|
| A. _Autosomes_ | A. _Meiotic_ |
| Trisomy | Deletion |
| Monosomy | Duplication |
| Triploidy | Translocation |
| | Ring Chromosomes |
| | Isochromosomes |
| B. _Sex Chromosomes_ | Inversions |
| Monosomy | B. _Mitotic_ |
| Trisomy | |
| Tetrasomy | Breakage |
| Pentasomy | Dicentrics |

characteristics of these individuals usually reveals some abnormality.

Some of the tissues of the human body are engaging in massive cell division and replacement each day. Over a long lifetime, therefore, opportunity is provided for the accumulation of many mitotic mutations in somatic cells, including both point mutations and chromosomal damage. Theories of carcinogenesis and of aging have been based on such mutations but, although appealing, such theories require additional tests(see Berenblum, this volume).

The above is a qualitative description of the effects of mutations. As for a quantitative estimate of hazards to human life and health, there are surprisingly few data. For the question, "How much genetic disease is there in the human population?", there is no answer. The best one can do is to estimate the proportion of genetically blighted persons in some circumscribed population. Even here, there are problems of ascertainment of all of the cases and of the definition of the population to serve as the denominator. Accordingly, we have asked the limited question, "How many patients admitted to two university hospitals during stated periods have diseases due to, or significantly influenced by, gene mutation or chromosomal aberration?"

Records were obtained for admissions to the Pediatric Medical Service of the Johns Hopkins Hospital for the period July 1, 1963 to June 30, 1969, and for the Adult Medical Service of the New York Hospital-Cornell Medical Center for the year 1968. Patients were counted only once, even though they might have been admitted several times in one year, and only the primary discharge diagnosis was used.

At Johns Hopkins there were 9352 patients admitted during the six-year period, and 3298 patients were admitted to the Medical Service at New York Hospital during 1968. Although collection and analysis of these two sets of data were similar in general, it is more convenient to discuss them separately.

Table II. Number and Frequency of Genetic and Non-Genetic Diseases in the Pediatric Service, Johns Hopkins Hospital, July 1, 1963 to June 30, 1969.

| Cause of Admission | Number of Cases | Frequency |
|---|---|---|
| Single Gene Defect | 597 | .064 |
| Chromosome Abnormality | 70 | .007 |
| TOTAL GENETIC | 667 | .071 |
| Possibly Gene Influenced | 2941 | .315 |
| Non-Genetic | 4982 | .532 |
| Unknown | 762 | .082 |
| TOTAL | 9352 | 1.000 |

Table II shows the division into categories of the 9352 pediatric patients. It was easy to decide which cases had disease due to single gene mutations or chromosomal abnormalities and which cases had non-genetic diseases. Together these categories constitute about 60 percent of the total. For the remaining 40 percent, there is no well-understood cause for each patient, although some disorders have more obvious genetic content than others. For example, diabetes is generally recognized to be a genetic disorder, although it is not due to a single gene. Some congenital malformations show family distributions which suggest single gene origin in a few families, and others, for example cleft palate or dislocation of the hip, are more frequent among relatives than in the general population, suggesting some multigenic inheritance. For other disorders there are insufficient data attesting to genetic origin, but the familial incidence is high enough to raise suspicion. Obesity and allergy fall into this subcategory.

Table III gives more detailed information on three of the classes. Among the "Non-Genetic" cases, infections were the largest component, comprising

Table III. Cases Listed as Possibly Gene Influenced, Non-Genetic, and Unknown Cause. Pediatric Service, Johns Hopkins Hospital, July 1, 1963 to June 30, 1969.

| POSSIBLY GENE INFLUENCED | | NON-GENETIC | |
|---|---|---|---|
| Allergy | 194 | Diarrhea and Dehydration | 372 |
| Autoimmune | 73 | Infections | 2201 |
| Cardiac Anomalies | 970 | Prematurity | 680 |
| Cataract and Glaucoma | 30 | Renal Disease | 495 |
| Central Nervous System Disorders | 22 | Trauma and Poisons | 462 |
| Epilepsy | 299 | Miscellaneous | 772 |
| Essential Hypertension | 22 | | |
| Hematologic | 52 | Total | 4982 |
| Malformations | 526 | | |
| Mental Retardation | 98 | | |
| Metabolic | 529 | UNKNOWN CAUSE | |
| Rheumatic Fever | 126 | | |
| | | Cardiac | 75 |
| Total | 2941 | Hernia | 123 |
| | | Neoplasms | 264 |
| | | Newborn | 184 |
| | | Strabismus | 32 |
| | | Miscellaneous | 84 |
| | | | |
| | | Total | 762 |

one-fourth of all admissions. (Although it is true that the mortality from infections has dropped remarkably, it is evident that hospital beds are still needed for their care.) Among the list of disorders headed "Possibly Gene Influenced", some patients may be included whose disease is in fact due to genes at a single locus, since the chart of each patient was not examined. The subcategories, however, are those which many would agree are likely to contain cases with some genetic influence. As an example, the "Metabolic" cases for the year July 1, 1968 to June 30, 1969 are given in detail in Table IV. Except for infections, the largest categories in Table III are malformations of the heart and of other organs. One simply doesn't know, or have any way of guessing, whether a few or many of the cases included under these headings are the result of new mutations. We may soon learn something about that, however, from patients with surgically repaired cardiac anomalies who have begun to have children. Certainly not all of the 299 cases of epilepsy are genetically determined, but there is evidence that some forms of epilepsy do have a genetic component. The list of cases of "Unknown Cause" is the most arbitrary of all. Little is known of the cause of any of these. The neoplasm category does not contain retinoblastoma

Table IV. Metabolic and Endocrine Cases Listed Among Cases Which May Be Gene Influenced. Pediatric Service, Johns Hopkins Hospital, July 1, 1968 to June 30, 1969.

| | |
|---|---|
| Diabetes Mellitus | 25 |
| Dwarfism | 36 |
| Hypoglycemia | 11 |
| Hypopituitarism | 21 |
| Obesity | 13 |
| Pubertal Variations | 6 |
| Thyroid Disease | 17 |
| Miscellaneous | 12 |
| Total | 141 |

or intestinal polyposis, conditions in which there is clear evidence of genetic origin.

Tables V and VI place the diseases due to a single mutant gene and the chromosome abnormalities in more precisely defined classes. Of these 666 cases, 227 (34 percent) can be dealt with by more than supportive treatment, while 73 (11 percent) are susceptible to prenatal diagnosis after amniocentesis. Thus 300 cases (45 percent) are partially or well controlled now, and we may be sure that this figure will rise rapidly in the future.

To summarize, one patient in 14 has a clear-cut genetic disorder, more than half do not, and the diseases of an additional nine percent seem very unlikely to have any genetical component. This leaves 30 percent of the cases in which it is difficult to decide. An arbitrary decision that one-fifth to one-half of these cases have some significant genetic background suggests that about 15 percent to 20 percent of patients admitted to this hospital during the period examined had diseases in which the genes played a large part.

It should be pointed out that these incidence figures are representative only of samples taken from university hospitals where there are clinical research units and eager investigators. They are not likely to be duplicated in community hospitals. An additional bias is due to the high proportion of the

Table V. Cases of Genetic Disease Due to Chromosome Abnormalities and to Single Gene Defects (Hematologic Disorders, Hereditary Anemias, and Inborn Errors of Metabolism). Pediatric Service, Johns Hopkins Hospital, July 1, 1963 to June 30, 1969.

| CHROMOSOME ABNORMALITIES | |
|---|---|
| Turner's Syndrome | 15 |
| Mongolism | 49 |
| Klinefelter's Syndrome | 1 |
| Other Trisomies | 5 |
| Total | 70 |
| **HEMATOLOGIC DISORDERS** | |
| Aldrich Syndrome | 1 |
| Dysgammaglobulinemia | 12 |
| Hemophilia A | 78 |
| Hemophilia B | 2 |
| Von Willebrand's Disease | 5 |
| Total | 98 |
| **HEREDITARY ANEMIAS** | |
| Diamond Blackfan Anemia | 3 |
| Fanconi Anemia | 1 |
| Hemoglobin Zurich | 1 |
| Hereditary Spherocytosis | 8 |
| S-Thalassemia | 1 |
| Sickle Cell Anemia | 99 |
| Thalassemia | 4 |
| Total | 117 |

| INBORN ERRORS OF METABOLISM | |
|---|---|
| Adrenogenital Syndrome | 52 |
| Albinism | 2 |
| Crigler-Najjar Syndrome | 1 |
| Cystinosis | 8 |
| Fabry's Disease | 1 |
| G6PD Deficiency | 4 |
| Galactosemia | 1 |
| Gangliosidosis | 1 |
| Gaucher's Disease | 4 |
| Glycogen Storage Type I | 1 |
| Glycogen Storage Type II | 1 |
| Glycogen Storage Type III | 2 |
| Goitrous Cretinism | 2 |
| Homocystinuria | 2 |
| Hyperlipemia | 2 |
| Hyperuricemia | 1 |
| Hypoglycemia | 5 |
| Lactase Deficiency | 2 |
| Metachromatic Leukodystrophy | 6 |
| Mucopolysaccharidosis | 5 |
| Oxalosis | 2 |
| Phenylketonuria | 16 |
| Porphyria | 2 |
| Pyruvate Kinase Deficiency | 1 |
| Tay-Sachs Disease | 7 |
| Wilson's Disease | 10 |
| Total | 141 |

Table VI. Miscellaneous Genetic Diseases Due to Single Gene Disorders Which Are Not Well Characterized Biochemically. Pediatric Service, Johns Hopkins Hospital, July 1, 1963 to June 30, 1969.

| Disease | Cases |
|---|---|
| Amyotonia Congenita | 2 |
| Chondrodystrophy | 16 |
| Cystic Fibrosis | 74 |
| Diabetes Insipidus | 3 |
| Dysautonomia | 4 |
| Dystonia Musculorum | 2 |
| Ectodermal Dysplasia | 2 |
| Ehlers-Danlos Syndrome | 1 |
| Familial Spastic Paraplegia | 1 |
| Fibrodysplasia Ossificans | 2 |
| Hereditary Nephritis | 26 |
| Hereditary Nephrosis | 1 |
| Hirschprung's Disease | 14 |
| Kartagener's Syndrome | 1 |
| Kugelberg-Wellander Syndrome | 1 |
| Lawrence-Moon-Biedl Syndrome | 2 |
| Male Pseudohermaphroditism | 9 |
| Marfan's Syndrome | 4 |
| Milroy's Disease | 1 |
| Morquio's Disease | 3 |
| Muscular Dystrophy | 24 |
| Myoclonus Epilepsy | 2 |
| Myotonic Dystrophy | 1 |
| Nephrogenic Diabetes Insipidus | 2 |
| Neurofibromatosis | 7 |
| Osteogenesis Imperfecta | 3 |
| Osteopetrosis | 1 |
| Polycystic Kidneys | 9 |
| Resistant Rickets | 6 |
| Retinitis Pigmentosa | 1 |
| Retinoblastoma | 5 |
| Tuberous Sclerosis | 5 |
| Werdnig-Hoffman's Disease | 5 |
| Total | 240 |

cases admitted to Johns Hopkins Hospital which are referred by physicians in Baltimore and other parts of Maryland. In addition, about 10 percent (254) of the cases for the year 1968-1969 came from outside the state (Table VII). Of these, 70 percent had one of the following: congenital heart lesions (26 percent), epilepsy (7 percent), endocrine disorders (17 percent), or single gene disease (20 percent); that is, most of the out-of-state patients had diseases which were overtly genetic or might be.

While similar morbidity data for cases at other hospitals were not found, comparisons can be made with previously published figures on mortality due to genetic and other disorders. These are given in Table VIII.

The data from the New York Hospital Medical Service are given in Tables IX and X, where they are presented in forms easily compared with Table II and Tables III and IV, respectively, showing data from Johns Hopkins. Major differences in incidence for each category in Table IX are immediately apparent. The frequency of patients in both "Single Gene Defect" and "Possibly Gene Influenced" categories are much lower than those for the pediatric cases, but still accounted for 10,445 days of patient care at a cost of approximately $1,300,000 and represented 11 percent of all care provided. Table X reveals that both the kinds and numbers of diseases differ markedly too. This is particularly obvious in the

Table VII. Distribution of Residence of Patients in the Pediatric Service, Johns Hopkins Hospital, July 1, 1968 to June 30, 1969.

| | Patients from Out-of-State | | Patients from Maryland | | |
|---|---|---|---|---|---|
| Category | Number | Percent of Category | Number | Percent of Category | Total Cases in Category |
| Chromosome | 4 | 19.0 | 17 | 81.0 | 21 |
| Hematologic | 2 | 3.2 | 61 | 96.7 | 63 |
| Inborn Errors | 14 | 40.0 | 21 | 60.0 | 35 |
| Miscellaneous | 30 | 40.0 | 45 | 60.0 | 75 |
| SUBTOTAL | 50 | 25.8 | 144 | 74.2 | 194 |
| Other Cases | 204 | 8.6 | 2162 | 91.4 | 2366 |
| ALL CASES | 254 | 9.9 | 2306 | 90.1 | 2560 |

Table VIII. A Comparison of Morbidity Rates at Johns Hopkins Hospital with Mortality Rates at Two Other Hospitals.

| | Hospital for Sick Children Great Ormond Street* | Newcastle upon Tyne# | Johns Hopkins |
|---|---|---|---|
| | Percent | Percent | Percent |
| Genetic | 12 | 11 | 7 |
| Gene Influenced | 25 | 31 | 32 |
| Non-Genetic | 15 | 41 | 53 |
| Unknown | 48 | 17 | 8 |

*Carter, 1956
#Roberts et al., 1970

Table IX. Number and Frequency of Genetic and Non-Genetic Disease Cases Admitted to the Medical Service of the New York Hospital, Cornell Medical Center, New York, 1968.

| Cause of Admission | Number of Cases | Frequency |
|---|---|---|
| Single Gene Defect | 49 | .015 |
| Chromosome Abnormality | 0 | .000 |
| Possibly Gene Influenced | 397 | .120 |
| Non-Genetic | 2,852 | .865 |
| Total | 3,298 | 1.000 |

lists of "Inborn Errors of Metabolism" and the "Possibly Gene Influenced" disorders where diabetes, essential hypertension, and their complications alone accounted for 78.5 percent. No doubt these differences are due to many factors, but notable among them must be the high mortality of many of the pediatric diseases, the satisfactory treatment of others, and the disposition to specialized hospitals, such as institutions for the mentally retarded, of still others. In addition, the late age of onset of the single gene diseases of adult life and of the degenerative disorders of middle life and old age preclude their appearance in a pediatric hospital. The latter reasons probably account for the differences in kind, the former for the differences in incidence, between the sets of figures.

Table X. Cases of Disorders Which Are Possibly Gene influenced and Due to Single Gene Defects. Medical Service, New York Hospital. Cornell Medical Center, 1968.

| POSSIBLY GENE INFLUENCED DISORDERS | |
|---|---|
| Asthma | 15 |
| Cancer Breast | 18 |
| Complications Diabetes | 153 |
| Complications Essential Hypertension | 136 |
| Diabetes | 30 |
| Essential Hypertension | 31 |
| Hyperlipidemias | 12 |
| Rheumatoid Arthritis | 2 |
| Total | 397 |

SINGLE GENE DISORDERS

| INBORN ERRORS OF METABOLISM | |
|---|---|
| Acute Intermittant Porphyria | 2 |
| Cystinuria | 1 |
| Lactase Deficiency | 3 |
| Morquio's Syndrome | 1 |
| Pseudohypoparathyroidism | 1 |
| Tyrosinemia | 1 |
| Subtotal | 9 |

| NEUROLOGICAL DISORDERS | |
|---|---|
| Dystonia Musculorum Deformans | 2 |
| Huntington's Chorea | 3 |
| Muscular Dystrophies | 6 |
| Spinal Degenerations | 4 |
| Subtotal | 15 |

| HEMATOLOGICAL DISORDERS | |
|---|---|
| G6PD Deficiency | 3 |
| Hemophilia A | 2 |
| Hereditary Spherocytosis | 1 |
| Sickle Cell Anemia | 5 |
| Thalassemia Major | 1 |
| Von Willebrand's Disease | 1 |
| Subtotal | 13 |

| MISCELLANEOUS SINGLE GENE DISORDERS | |
|---|---|
| Familial Myocardopathy | 1 |
| Hemachromatosis | 2 |
| Marfan's Syndrome | 2 |
| Osler Weber Rendu | 2 |
| Polycystic Kidney | 4 |
| Subtotal | 11 |
| Total | 48 |

All of these data seem to be telling us that, while many patients come to the hospital with diseases due to genes at a single locus, many more, perhaps twice as many, have disorders in which there is some evidence of partial genetic cause, but one knows nothing about the numbers and kinds of genes involved. Evidently a large amount of genetically-determined disease remains to be characterized. Many of the multigenic disorders, although milder in their effects and sometimes manifested only in adulthood, are very common, much more so than the inborn errors of metabolism, most of which have frequencies of one in 25,000 or less. These multigenic disorders are difficult even to characterize as genetic in their origin, let alone to discover the mutant genes and describe their action. Clearly some new ways of thinking about multigenic variation are needed.

## References

Carr, D. H. (1969). Prog. Med. Genetics, 6, 1.
Carter, C. O. (1956). Great Ormond Street Journal, 11, 65.
Lewontin, R. C. (1967). Ann. Rev. Genetics, 1, 37.
McKusick, V. A. (1970). Ann. Rev. Genetics, 4, 1.
McKusick, V. A. (1968). Mendelian Inheritance in Man, 2nd edition. Baltimore, Johns Hopkins Press.
Ratcliffe, S. G., A. L. Stewart, M. M. Melville, P. A. Jacobs, and A. J. Keay (1970). Lancet, 1, 121.
Roberts, D. F., J. Chaney, and D. M. Court (1970). Arch. Dis. Child., 45, 33.

# THE MOLECULAR BASIS OF MUTATION*

*John W. Drake and W. Gary Flamm*

To assess the potential dangers of environmental mutagens, one must first understand the mutation process, including the nature and origins of mutations and the ways in which mutations are likely to be produced by various agents that impinge upon man. One must also have some understanding of mechanisms of inheritance and the role of mutation in evolutionary development.

A normal human being possesses 22 homologous pairs of chromosomes (autosomes), one of paternal and the other of maternal origin, as well as a sex chromosome from each parent. Each homologous pair of chromosomes, however, generally has numerous point-for-point differences. Most genes in a population, for instance, are present in several slightly different versions, probably because some unlike pairs of genes may work better than an identical pair and also because different versions of a gene may be more successful in different local environments. The source of this variability is mutation. However, mutations arise in a completely random fashion, and in a complex and finely tuned organism such as man, mutations cause mainly deleterious effects.

---

*Most of the information which is summarized here has been presented much more extensively in *The Molecular Basis of Mutation* by John W. Drake (San Francisco, Holden-Day, 1970) and in *Chemical Mutagens, Environmental Effects on Biological Systems* by L. Fishbein, W. G. Flamm, and H. L. Falk (New York, Academic Press, 1970).

## The Genetic Material

All mammals (and most other animals as well) begin life in the same way--as a fertilized egg. The differences among them lie in the genetic material, the substance that has been called the blueprint of life. In chemical terms, the material is deoxyribonucleic acid (DNA), a substance of enormous length in animal cells and about ten million times longer than its diameter. In fact, if the DNA contained in one cell were stretched out as a single piece, it would measure approximately three feet. Nature has managed to package this DNA so that it fits into a cell only about 0.001 inch in diameter. There are about 20 trillion cells in an adult person, and if the DNA of these cells were laid end-to-end, it would stretch from the earth to the sun and back again 50 times over. This comparison is not entirely fair, since each of the 20 trillion cells contain identical three-foot pieces of DNA and, hence, essentially the same genetic information. Each of the DNA's in the trillions of cells originates from the three-foot piece contained in the fertilized egg. Human chromosomes contain enough DNA to make up well over a million genes, although the bulk of this DNA may not be genes and its functions remain obscure.

## Replication of Genetic Material

An understanding of the structure of DNA and of the way trillions of copies of DNA are made to accommodate trillions of cells is essential to understanding mutation. DNA is a double-stranded molecule and is composed of what appears to be two thin ropes (strands) which are tightly twisted about each other. There are four main constituents in DNA, called deoxyadenylic acid (A), deoxythymidylic acid (T), deoxyguanylic acid (G), and deoxycytidylic acid (C). These four constituents, called nucleotides, are linked together linearly to form the DNA. These constituents are able to recognize each other through hydrogen bonding: A and T specifically recognize and pair with each other due to hydrogen bonds, while G and C represent the other partnership. All of the A's, T's, G's, and C's on one strand of the DNA are opposite and paired with the T's, A's, C's, and G's, respectively, on the other strand. Consequently, each strand represents a template or blueprint for the synthesis of its partner strand (or complementary strand, as it is more generally called).

Before a cell divides to produce two daughter cells, the two strands of its DNA separate, and each serves as a template for synthesis of a new strand. Since the A's, T's, G's, and C's pair only with their respective partners, the two new double-stranded DNA's are duplicates of the original DNA prior to strand separation. In other words, each strand has specified the synthesis of its complementary strand because, due to the specificity of hydrogen-bonding, it could not have done anything else. Hence, before a cell divides into two daughter cells, its DNA has already duplicated by undergoing strand separation and replication. Upon cell division, one pair of DNA strands (one old strand and one new complementary strand) accompanies one daughter cell, while the other duplex remains with the second daughter.

## Utilization of Genetic Information

If one of the A's, G's, C's, or T's fails to pair with its proper partner, an incorrectly paired substituent can be incorporated into the DNA during replication. This mistake constitutes a mutation. If the mistake is not lethal to the cell, the mutation will continue to be passed down in subsequent DNA replications. This mutational alteration is manifested when the genetic information is utilized by the cell to direct the cell what to do and what to become. The first step in utilizing genetic information is called "decoding" or "reading of the gene."

A single gene contains, on the average, about 1,000 nucleotides (A, G, T, and C) arranged in a specific linear order or sequence. This group of ordered nucleotides is divided into sets of three's, called "triplets" or "code words." Each triplet specifies an amino acid, of which there are about twenty, and the order of the three nucleotides in the triplet determines which amino acid. For instance, the order of a triplet containing two G's and one C might be GGC, GCG, or CGG. Each of these specifies a different amino acid.

The twenty amino acids are the constituents of which proteins are composed. A protein is a linear sequence of different types of amino acids linked together, similarly to the way the nucleotides of DNA are linked together linearly. Proteins carry out the cell's work and give the cell its shape and form. They provide the means by which foodstuffs are utilized for the release of energy needed for all life processes. Proteins also serve as structural components, providing shape and form not only to the individual cells but to the organism as well. Clearly, since DNA contains the code for protein synthesis, it possesses the key to life.

The series of code words required to specify a complete protein molecule is the equivalent of a gene. Each gene occupies a specific part of the DNA of a chromosome and the average length of a gene is only one-ten millionth of the three foot piece, or about one-ten thousandth of an inch. In a single human cell, there is enough DNA for about ten million such genes. The actual number, however, is in doubt, since not all the sequences of DNA serve to specify proteins.

This relatively small stretch of DNA (the structural gene) does not act directly as a coded template for production of specific proteins. Instead, it acts as a template for the synthesis of a molecule which is chemically very much like itself and is about the same length as the gene. The molecule is aptly named messenger RNA (ribonucleic acid) and contains A, G, C, a derivative of T known as uracil (U), and ribose instead of deoxyribose. RNA is synthesized from a DNA template in much the same way that DNA itself is synthesized, i.e., complementary copies are synthesized from template strands. This messenger molecule then migrates to the cellular sites where proteins are made. There it directs the synthesis of protein through the use of the triplet code words (codons) which, of course, it obtained from its complementary template DNA.

The codons in messenger RNA are specifically recognized during protein synthesis by "anticodons" (the triplets complementary to codons) of so-called transfer RNA molecules. These molecules have attached to them specific amino acids and contain a single, specific anticodon, e.g., the anticodon on the transfer RNA which carries the amino acid alanine is CGG which recognizes, through hydrogen bonding, the complementary codon GCC for alanine on the messenger RNA.

To summarize briefly, DNA replicates via a process involving strand separation and the use of individual strands as templates for the synthesis of complementary copies. One strand of DNA serves as a template for the synthesis of messenger RNA, which directs the amino acid sequence of proteins. This is accomplished by the sequence of triplet code words (codons) on the messenger RNA which are complementary to the triplet anticodons on the transfer RNA's carrying the amino acids.

## The Nature of Mutations

Mutation is the process by which mutants are formed, and a mutant is an offspring or individual that is genetically unlike its parent because of sudden fundamental change in its hereditary material. Since the nature of genetic material is known, we can go further and define a mutant in terms of the molecular events responsible for it.

Mutations can be divided into large and small, according to whether they affect an appreciable length of the chromosome or only one or a few points along its length. The large mutations consist either of changes in the number of chromosomes in a cell, or of large changes within and among the chromosomes. The loss of one or more chromosomes is very apt to produce drastic effects, and typically results in the death of the embryo at a very early stage resulting, for example, in spontaneous abortion. The gain of an additional chromosome, as in mongolism, also typically produces a drastic effect, since the balanced interplay among many thousands of genes is disturbed.

Changes within and among the chromosomes consist both of additions and of deletions of DNA sequences and also of rearrangements of long sequences without any net change of DNA content. Deletions and additions may produce effects as drastic as do chromsome gains and losses, depending upon the length and nature of the region affected. Rearrangements, which consist either of inversions (in which a segment is reversed within a chromosome) or of translocations (in which a segment is moved to a new location), produce highly variable effects.

The small mutations, called point mutations, consist of base pair substitutions and of base pair deletions and additions. The effects produced by point mutations vary greatly, depending both upon the type of mutation and upon the affected gene.

If, by means which will be discussed later, one of the triplet code words in the gene is changed, messenger RNA would likewise be affected and an incorrect amino acid might be specified by the abnormal triplet. This is considered a mutation and, if the protein's function were altered, the mutant might be readily noticeable. This type of mutation often produces what is called a missense mutant, meaning that a code word which specifies one amino acid has been changed to a different code word which specifies a different amino acid. For example, the original code word might have been AAA, which specifies the amino acid lysine. If this is somehow changed by mutation to GAA, glutamic acid would be incorporated into the protein in place of lysine.

Substition does not necessarily severely alter the affected protein. The severity of the effect depends upon where it occurred in the amino acid sequence and the nature of the substitution (i.e., which nucleotide was substituted for which). If a critical region of the protein is involved, there may result a complete loss of function. More frequently, the substitution occurs in a relatively non-critical region of the molecule, allowing the mutant protein to carry out its function as before but with, perhaps, reduced ability.

It is also possible that the altered protein will possess the same functional abilities it had prior to the mutational event, and these are called neutral mutations. Some mutations involve a change in one of the nucleotides of a code word and yet do not change the code itself. Mutations of this type are called silent. For instance, the amino acid alanine can be specified by either GCC, GCA, or GCG, and, consequently, changes in the last place of this code word are not likely to lead to changes in amino acid coding.

In certain rare cases, a mutation may produce an advantageous change, and it is through such mutations that organisms and species evolved to give rise to the great variety of life found on this planet. However, it should be realized that the vast majority of the mutations which occur in nature are harmful insofar as the individual is concerned.

Besides generating missense mutants, substitution of one nucleotide for another can also result in the development of a nonsense code word. This is a code word which is unable to specify any amino acid at all, and normally these are used to terminate protein synthesis when the protein is complete. However, when they occur in the middle of a genetic message as a consequence of mutation, protein synthesis is prematurely terminated at the nonsense mutation. Nonsense mutations generally have a far more serious effect on the gene product (the protein) than mutations which give rise to missense mutants.

Besides mutations which involve substitutions of one nucleotide for another, mutations are also caused by the addition or deletion of nucleotides. These give rise to frameshift mutants, an excellent name, since it succinctly describes what happens. The code words of a gene are in tandem, with no separation. Hence, if one nucleotide is added or removed, the entire message loses its meaning. For instance, if we write the message HOW CAN THE RAT EAT, but remove the spaces and let the reader know that each three letters in a series form a word starting with H, we have HOWCANTHERATEAT. If the first T is deleted, the message is changed to HOWCANHERATEAT, which affects all the code words on the right of the deletion and, hence, the wrong amino acids would be specified.

Clearly, either a deletion or an addition of a nucleotide would lead to a scrambled message, and since many code words are affected, frameshift mutations generally destroy totally the activity of the gene in which they reside. Most base pair substitutions, on the other hand, produce relatively mild effects upon the activity of the affected gene.

## Mechanisms of Mutagenesis

It is now fairly clear how mutations arise. The large mutations result from errors of chromosome distribution among progeny cells and from unrepaired or misjoined chromosome breaks. Point mutations arise in two rather different ways. They may arise from errors occurring during duplication of the chromosome before cell division, in which either a normal base or a chemically modified base mispairs to produce a base pair substitution. They may also arise from errors of repair, in which a region of damaged DNA is removed and resynthesized. The cell produces a variety of repair sytems for detecting and eliminating errors in DNA, but some of these sytems are rather inaccurate. They are able to remove most potentially lethal damages, but occasionally they introduce an error which is then perpetuated as a mutation. Errors of repair may be both base pair substitutions and frameshift mutations; the latter, in fact, are probably produced almost exclusively by misrepair.

We are primarily concerned with mutations produced by factors extrinsic to the body. A very large number of agents, including numerous important chemicals, have been shown to induce mutations in lower organisms in laboratory tests. While it is not yet clear which of these agents may actually cause mutations in man, the potential human cost of a misjudgement of any considerable magnitude is very great. As a result, any agent which produces mutations in laboratory test systems must be considered with grave suspicion. It is now reasonably clear that both the chemical structure and the function of DNA is very similar in simple and in complex organisms. The unity of genetic mechanisms provides the fundamental justification for extrapolating from various test systems to man. Although man may protect himself against mutation, for instance by elaborating repair and detoxification processes, he certainly does not do so with perfection, nor is he likely to do so in just the way as does, for instance, a mouse or a fruit fly.

The agents (mutagens) that produce mutations in test systems act by only a few basic mechanisms. They may react directly with DNA. They may react with or replace one of the components that will later be built into DNA. They may interfere with the availability of the components needed for DNA synthesis. Or they may react with proteins which in turn cause mutations in the DNA, either by failing to function at all or by functioning abnormally.

Many mutagens react chemically with DNA bases. Nitrous acid, which is produced when the preservative sodium nitrite meets stomach acids, reacts with cytosine and with adenine to produce modified bases. These then engage in the unnatural pairings C*:A and A*:C in the next DNA replication. Two other chemicals which react with cytosine to produce C*:A mispairs are sodium bisulfite, a common preservative in foods and pharmaceuticals, and hydrazine, a rocket fuel.

Many other chemicals do not cause specific mispairing but do activate repair systems, whereupon misrepair becomes possible. Much of the mutagenicity of ultraviolet irradiation, for example, occurs in this manner. Many alkylating agents and nitrosamines are mutagenic, and they probably act partly by triggering misrepair. Nicks in the DNA backbone often lead to frameshift mutations, and both ionizing radiations and many chemicals cause such backbone nicks. In addition, many chemicals act by interfering with the repair process itself rather than merely by triggering it. One striking example is atabrine, which was widely used as an antimalarial drug.

A few mutagenic chemicals resemble DNA bases and are by mistake incorporated into DNA. However, these base analogs exhibit equivocal pairing properties, acting at one replication like one base and at the next replication like another base. Some antibiotics and many anticancer drugs are base analogues and potential mutagens.

Depriving an organism of DNA bases can be markedly mutagenic. A variety of agents interfere with the synthesis of the bases or with their delivery to the site of DNA synthesis. Sometimes these agents act as analogues of base precursors, and sometimes they act specifically to inactivate one of the enzymes involved in base synthesis. Again, some antibiotics and anticancer drugs act in this manner. It is also possible that grossly inadequate or unbalanced nutrition is mutagenic, although this possibility is insufficiently explored at present.

It will be well to end this brief survey of mutagenesis by noting the systems which are employed by organisms to avoid mutagenesis. In the first place, the enzymes which carry out DNA synthesis have evolved a certain selectivity and are often able to reject an incorrect base or base pairing. Next, of course, come the repair enzymes. The well characterized repair systems are directed primarily at avoiding lethality, however, and repair itself is frequently sufficiently inaccurate to introduce new mutations.

The next important mechanism for avoiding mutagenesis is detoxification, the destruction of noxious agents by cells and tissues. Like repair, however, detoxification can be a two-edged sword; certain compounds which are not themselves mutagenic are converted, in the course of their breakdown, into mutagenic derivatives. An example is the conversion of cyclamate into cyclohexylamine.

At the highest level, organisms collectively avoid the effects of mutation by natural selection. Man, however, often struggles to mitigate selection and is likely to continue to do so. Selection nevertheless continues to operate and must operate even more strongly if the human mutation rate increases.

## The Effects of Mutation on the Organism

Since chromosomes occur in pairs (except for the special case of eggs, sperm, and sex chromosomes), it might be expected that a mutationally damaged gene on one of the chromosomes could be compensated by a normal gene on the other chromosome. This is very often true; the mutation is displayed only when carried by both of the chromosomes. Such a mutation is said to be recessive, and its opposite type, a mutation which is clearly expressed even in the presence of a normal homologous gene, is said to be dominant. Recessive mutations are rather difficult to detect in natural populations such as man, since the likelihood of a new mutant gene combining with a similar mutant, either new or old, is very small in any one generation. Dominant mutations are expressed immediately and therefore have been more extensively studied.

The significance of mutations in humans, of course, is that they frequently manifest themselves as clinical diseases. Most of the mutations which arise are very probably--certainly in the case of the fruit fly--only mildly deleterious. For instance, they might only be detected as a reduction of a few percent in the average number of offspring produced. Just because they are so mild, however, these deleterious mutations persist much more successfully than would a truly drastic mutation. Over many generations, therefore, their cumulative action can produce an effect which is just as serious as that produced by a more immediately deleterious mutation. A second factor also affects the long-term cost accounting of the effects of mutations on a population. When the mildly deleterious mutations are combined together in single individuals, their effects may be more than simply additive. The more such mutations are present, the greater is the effect of adding one more mild mutation. Although the average load of such mild mutations in man is not at all clear, the average fruit fly carries enough of them so that, for example, its potential for reproduction is only about 70 percent of optimal. Furthermore, a doubling of the mutation rate in the fruit fly would produce an additional lowering of vitality of a good deal more than the present differential of 30 percent.

Some genes can be totally inactivated by mutation without an immediate and obvious effect upon the organism, even when both copies of the gene are affected. These genes are apparently dispensable under the particular conditions of observation, but are nevertheless likely to become very important under some other set of conditions. Even though an organism may not encounter a certain critical environment very often, his genetic potential is constructed to allow him to survive in a wide range of different environments. The proportion of superficially dispensable genes in man is very difficult to estimate, but could be quite large. An accumulation of mutations in these genes could be very dangerous, since man's ability to withstand changing conditions could be weakened. There is, in fact, some evidence to indicate that the human environment is changing more rapidly now than in past recorded history. While mutation is not the only process by which such agents interfere with normal viability and reproduction, its potential for harm is nevertheless immense and should be delineated as precisely as possible.

# MONITORING OF CHEMICAL MUTAGENS IN OUR ENVIRONMENT*

*Heinrich V. Malling*

## Introduction

A great number of chemicals are introduced into the environment by a modern technological society. In addition, these compounds can be converted in nature to other compounds, such as methyl mercury or nitrosamines. The identification of chemicals in our environment that have mutagenic activity has been increasing steadily, both through the efforts of a growing number of scientists working in this area and as a result of the development of various sensitive assay systems. Mutagenic activity is not confined to exotic laboratory chemicals. Chemical mutagens have also been found among drugs, pesticides, food additives, and cosmetic products, some of which have widespread use by the human population (see Fishbein and Nichols, this volume; Fishbein et al., 1970).

The genetic material in man and most other organisms is DNA, and the reactions between mutagens and DNA in man should be comparable to those in other organisms. Since there are many compounds in the environment which are mutagenically active in specific test organisms, it is likely that some of them can also induce mutations in man. Most mutations have a deleterious effect, and to surround the human population with chemical mutagens is to endanger health. The problem is with us now, since we are living in a population that consists of at least three generations that have been exposed to a wide variety of man-made chemical agents.

---

*Sponsored by the U. S. Atomic Energy Commission under contract with the Union Carbide Corporation.

How can we cope with this problem? To prevent more damage to the human population and to inhibit the increase of mutagenic chemicals in the environment, compounds should be tested for mutagenicity before they are released to use. The routine screening procedures now available for mutagenicity testing are new, and presently there is no requirement by the public health department of any developed country that products be tested before they are marketed. As a result, only a few of the chemicals now in daily use have been tested for genetic effects.

The compounds which need testing are numerous. Some are already known to be carcinogenic, teratogenic, and/or mutagenic, but most compounds now in use have never been tested for adverse biological effects other than lethality. Since the number of chemicals which should be tested for mutagenicity is large, it is necessary to establish priorities as to which ones need to be tested first. Priorities can be based on several criteria: (1) To what degree is the human population exposed to the particular chemical? (2) What is known already about its genetic effects? (3) Is there circumstantial evidence of biological effects which might indicate mutagenic activity? (4) Are there closely related chemicals which are mutagenic? (5) Does it have metabolites which might be mutagenic?

Testing should first be initiated for those which are most likely to be mutagenic, a decision which can be made after a meaningful assessment of the data in the literature on chemical mutagenesis, carcinogenesis, and teratogenesis. An initial step to collect this literature was taken by the Environmental Mutagen Society, which set as one of its main tasks the formation of the Environmental Mutagen Information Center (EMIC), officially established in October, 1969. The EMIC has adapted a computer program to make bibliographic information accessible for searching.

A total of 5,000 items have been entered into the data bank. From these data, six different bibliographic compilations in specific areas of interest have been assembled. (See Table I for EMIC publication list.) The EMIC has established

TABLE I. DATES AND TITLES OF PUBLICATIONS FROM EMIC

---

Adler, I. D., J. S. Wassom, and H. V. Malling (1971). A bibliography on the genetic effects of caffeine. Environmental Mutagen Society Newsletter, 4, 44.

Generoso, W. M. (1969). Literature citations on chemical mutagenesis in mammals. Environmental Mutagen Information Center Pamphlet #1.

Legator, M. S., J. S. Wassom, and H. V. Malling (1970). General literature collection on cyclamates. Environmental Mutagen Information Center Pamphlet #3.

Malling, H. V., and J. S. Wassom (1969). Literature survey on pesticides with special reference to their mutagenic activity. Environmental Mutagen Society Newsletter, 2, 34.

Malling, H. V., and J. S. Wassom (1969). Environmental Mutagen Information Center (EMIC). I. Initial organization. Environmental Mutagen Society Newsletter, 1, 16.

Malling, H. V., and J. S. Wassom (1969). Tabulation of the mutagenic effect of certain pesticides with supporting bibliographical references. Report of the Secretary's Commission on Pesticides and Their Relationship to Environmental Health. Parts I and II, U. S. Dept. of Health, Education, and Welfare, p. 611.

Malling, H. V., J. S. Wassom, and E. S. Von Halle (1970). A survey of the 1969 literature on chemical mutagenesis. Environmental Mutagen Society Newsletter, 3 (Suppl. 1), 1.

Malling, H. V., J. S. Wassom, and S. S. Epstein (1970). Mercury in Our Environment. Environmental Mutagen Society Newsletter, 3, 7.

Malling, H. V. (1971). Environmental Mutagen Information Center (EMIC). II. Development for the future. Environmental Mutagen Society Newsletter, 4, 11.

Wassom, J. S., E. Zeiger, and H. V. Malling (1970). A bibliography on the mutagenicity of nitroso compounds and related chemicals. Environmental Mutagen Information Center Pamphlet #5.

Wassom, J. S., and H. V. Malling (1971). The Environmental Mutagen Information Center during FY 1971. (Abstract of a paper delivered at the 2nd Annual Meeting of the Environmental Mutagen Society, Washington, D.C., March 21-24).

Wassom, J. S., and H. V. Malling (1969). Literature citations on chemical mutagenesis, January 1 to April 30, 1969. Environmental Mutagen Information Center Pamphlet #2.

Wassom, J. S., H. V. Malling, and F. J. de Serres (1970). Literature citations on drugs of abuse. 1. The genetic effects of LSD and related psychotic compounds. Environmental Mutagen Information Center Pamphlet #4.

contacts at various locations in the United States and at several universities and research institutions in Sweden and Germany, with scientists who have expressed their interest and their willingness to collaborate in a data collection program. In such a program, scientists would report their results on a standardized form, which could then be processed directly by the computer. The data will be organized as in Table II.

To increase understanding of chemical-biological interactions, implementation of a substructural computer search program would be extremely helpful. This would enable us to draw correlations between active groups and mutagenic activity, a capability that might have some prognostic value for predicting the mutagenic activity of newly created chemicals. Such an ability could provide financial savings and health protection through the creation of safer compounds for widespread use. Many scientific laboratories have tested pesticides, food additives, and other compounds now in widespread use in the environment for mutagenic activity in various testing systems and have obtained negative results. These results are often never published, but it is extremely important to get such results into the published literature in an organized and readily accessible form.

## Problems in Testing for Mutagenic Activity

As soon as priority lists of chemicals have begun to be established, the tremendous job of testing can be carried out. Three different test systems have been suggested: the induction of dominant lethal mutations, the induction of chromosome aberrations *in vivo*, and the host-mediated assay (Mrak, 1969). The host-mediated assay uses microorganisms and is probably the simplest of the three test systems. There are, however, several problems involved in the use of microorganisms, in the extrapolation to man of data obtained by these tests, and in the evaluation of the data from the test systems.

Table II. Proposed Tabulation of Mutagenesis Data

| Pesticide | Organism in which tested | Assay system | Dose | Biological effect | EMIC registry No. |
|---|---|---|---|---|---|
| 2,4-D | Tradescantia | | 0.001% to 1.0% (Minimum effective dose - 0.001%) | Abnormal mitosis | 60 |
| DCNA | Barley | Anther | 1,000 p.p.m.--Soaked | Slight effect on meiosis ($C_1$) | 70 |
| | | | 500 p.p.m.--Sprayed | High abnormal meiosis ($C_2$) | 70 |
| DDT | Mouse | Sperm | 105 mg/kg | No increase in frequency of dominant lethals | 23 |
| DDT | Allium cepa | Root tips | Saturated solution | C-mitosis and chromosome breaks | 408 |
| DDT | Trigonella | Root tips | Saturated solution | C-mitosis and chromosome breaks | 408 |
| Dichlorvos | Onion | Root tips | 0.5 to 6.0 sq. cm. | Chromosome breaks | 396 |
| Dicomba | Barley | Anther | 1,000 p.p.m.--Soaked | Abnormal meiosis ($C_1$) | 70 |
| | | | 500 p.p.m.--Sprayed | Abnormal meiosis ($C_2$) | 70 |
| Dieldrin | Crepis capillaris | Sprouts | 10% solution | C-mitosis effect, no chromosome breaks observed | 40 |
| Endothall | | Plant cells | | Chromosome aberrations | 570 |
| Endrin | Barley | Anther | 1,000 p.p.m.--Soaked | No effect on meiosis ($C_1$) | 70 |
| | | | 500 p.p.m.--Sprayed | No effect on meiosis ($C_2$) | 70 |
| Ethylene oxide | Fungi | | 0.025 M | Point mutations and reverse mutations | 258 |
| Ethylene oxide | Neurospora crassa | Conidia | 0.14 M | Point mutations and reverse mutations | 34 |
| Ethylene oxide | Maize | Plant cells | 1 part E.O. to 20 parts air | Chromosome breaks | 25 |
| Ethylmercury chloride | Triticum | Root tips | 0.5 to 1% | Mitotic aberrations | 357 |
| Ethylmercury chloride | Secale cereale | Root tips | 0.5 to 1% | Mitotic aberrations | 357 |

## 1. COMPARATIVE MUTAGENESIS

The most sensitive tests of mutagenicity employ microorganisms, and there are a number of chemicals which are clearly mutagenic in these organisms, but whose effect on higher organisms, and man in particular, is unknown. Nevertheless, since the composition of the genetic material (DNA) is the same in most organisms, a chemical which is mutagenic in one species is likely to be mutagenic in others and must be viewed with suspicion if it has a widespread use in the human population.

A direct test of this assumption can be made by comparative mutagenic studies of compounds in microorganisms and mammals under conditions which are as nearly identical as possible. For triethylene-melamine (TEM), methyl methanesulfonate (MMS), and ethyl methanesulfonate (EMS), enough data exist in the literature to compare some properties of these compounds--their reactions with DNA and their mutagenic effects in _Neurospora_ and in mice. TEM reacts with DNA and links two DNA strands together (Lorkiewicz and Szybalski, 1961). When cross-links occur, DNA is most likely not replicated at that point, and a chromosome break is the probable result. TEM induces a high number of dominant lethals in mice (Cattanach, 1959), which are often the result of chromosome breakage. TEM also induces chromosome deletions in _Neurospora_, as well as point mutations and nuclear killing (Malling and de Serres, 1969). In _Neurospora_, both EMS and MMS are less efficient than TEM in producing chromosome deletions. MMS and EMS both alkylate the nucleotides in DNA. When deoxyadenylic and deoxyguanylic acids are alkylated, they tend to be deleted from DNA (Freese, 1961), which can induce cross-linkage in DNA (Freese and Cashel, 1964). Since methylated nucleotides have a greater tendency to be deleted than ethylated bases, MMS should induce more cross-links than EMS. MMS induces a higher frequency of dominant lethals in mice than EMS (Ehling, Cumming, and Malling, 1968), indicating that formation of cross-links in DNA and induction of dominant lethals are positively correlated. However, it is difficult to reach general conclusions about mammals, since there are sex differences and strain differences in the induction of dominant lethals with these chemicals

(Generoso and Russell, 1969; Generoso, Huff, and Stout, 1971).

## 2. EVALUATION OF RESULTS FROM MUTAGENESIS TESTING

Lack of more complete knowledge in the field of comparative mutagenesis hinders the extrapolation of data from test organisms to man. An added difficulty is the fact that some positive and some negative results undoubtedly will be false. It is likely that public health departments in most countries will be worried more about false negative results, and that industries which are trying to market new products will be concerned about false positive results. It is of utmost importance that policy-making guidelines be established and that decisions be made on a scientific level in individual cases. Many points of discussion about test results will have to be raised, for example the following.

### a. Submammalian versus Mammalian Testing

Microorganisms can sometimes survive treatment with much higher concentrations of certain chemicals, e.g., hydroxylamine, than mammals (Malling, 1966a; Somers and Hsu, 1962). On the other hand, mammals possess strong detoxification mechanisms for certain chemicals, e.g., N-methyl-N'-nitro-N-nitroso-guanidine, and can be treated with much higher concentrations than microorganisms (Malling and Cosgrove, 1969). If a certain compound has given a positive result in any of the submammalian systems and not in whole animals, that compound should still be viewed with suspicion and not be used widely until the basis for the discrepancy between the two test conditions is clarified. The suspicion may be lifted only if its mutagenicity in microorganisms is due to certain biological mechanisms in these organisms which are absent in mammals.

### b. Reverse versus Forward Mutation Systems

The mutagenicity of a chemical is often more easily tested by detecting reversion of a mutation back to the normal gene (reverse mutation) rather than by detecting mutation in the opposite direction (forward mutation). There exist, however, some built-in errors. In reverse mutation, it is likely that only a few codons are the sites for the mutagenic action that is detected, and this might also be the case for certain forward mutation systems. Through a proper selection of mutants, most known point mutation events can be covered (Ames, 1971; Malling, 1967). However, mutants with similar genetic alterations at different sites in their genes revert with greatly different frequencies after treatment with the same mutagen (Benzer, 1961; Malling and de Serres, 1968). This might indicate that the molecular environment of a particular nucleotide has a great influence on the mutability of that nucleotide with a chemical mutagen. Moreover, point mutations can be caused by at least eight different mechanisms, only one or two of which will result in reverse mutations in a certain mutant. If detection of reverse mutations is a part of a general screening program, at least two mutants of each type of reverse mutation mechanism should be included in the test. There is no doubt that a positive result in a reverse mutation system indicates mutagenicity. A negative result, unfortunately, can be interpreted in two different ways: either the chemical is not mutagenic under the test conditions, or the active chemical cannot react with the DNA at the site that would result in a reverse mutation, in which case the negative result is false.

There are forward mutation systems, however, in which alterations at many different sites can be scored as mutations, e.g., the purple adenine system in _Neurospora_ (de Serres and Osterbind, 1962; de Serres and Malling, 1971; de Serres, this volume). A similar and promising system based on canavanine resistance in yeast (Brusick, 1971) is now being developed. In the _Neurospora_ system, both transmissible chromosome deletions and point mutations are scored simultaneously; in the yeast system only point mutations are scored. A negative result in this type of forward mutation system is likely to be more meaningful than a negative result in a reverse mutation system.

### c. Influence of Mammalian Metabolism

Mammals are able to detoxify strong mutagens and to form mutagenic metabolites from nonmutagenic chemicals. The test for mutagenicity of known or unknown metabolites can be carried out in at least three different systems: The host-mediated assay (Gabridge and Legator, 1968); enzyme formation of the active metabolites *in vitro* and incubation of the indicator organism in the reaction mixture (Malling et al., 1971); and formation of the metabolites in chemical systems which mimic the metabolism in mammals (Malling, 1966b).

False negative results can easily occur in all these test systems. In both the host-mediated assay system and the enzyme system, the indicator organism is subjected to the chemical and its metabolites over a relatively short time period. In the host-mediated assay system, 3 hours incubation time is standard for *Salmonella typhimurium*, 18 hours for *Neurospora crassa*, and 12 hours for tissue cultures of Chinese hamster cells.

A chemical such as DDT, which accumulates in mammals and is slowly metabolized, may give rise to such minute amounts of mutagenic metabolites during the short duration of these tests that they would not be detected. Since human germinal cells are exposed to DDT and similar compounds from conception to the end of reproductive age, even small concentrations of mutagenically active metabolites could be of significant importance for the induction of mutations that would be inherited by the progeny of the exposed individual.

Testing chemicals for formation of mutagenic metabolites with enzymes from different human organs should be possible. False negative results are likely, however, due to the lack of required co-enzymes or other important factors, some of which may still be unknown. False positive results are also possible. For example, mutagenically active compounds might be formed by chemical hydroxylation which would not be formed during *in vivo* metabolism.

#### d. Sensitivity of Different Test Systems

Evaluation of the sensitivity of different tests is extremely difficult. For example, if sensitivity is expressed as mutations per ethylated nucleotide (Sega and Lee, 1970), then

*Drosophila* is probably much less sensitive than the mouse (Cumming and Walton, 1971; Russell and Cumming, 1971). However, X-irradiation is considerably more effective in inducing mutations in the 7-specific locus system in mice than in the purple-adenine system in *Neurospora* (Russell, 1951, 1967; Webber and de Serres, 1965).

## 3. WEAK MUTAGENIC ACTIVITY: A MAJOR PROBLEM AREA

Compounds which have widespread use by the human population and which have a weak mutagenic effect may be more important than strong mutagens to which only a limited number of people are exposed. Most of the mutagenic compounds in our environment probably have only weak mutagenic activity. To demonstrate weak mutagenicity is difficult and attempts to do so can lead to controversy between scientists and legislators. The study of the mutagenicity of caffeine can serve as an example.

The consumption of caffeine by humans is about 500 mg per person per day in individuals over 14 years of age. Most of the caffeine comes from coffee and tea, but some comes from cola and medicine. The degradation rate is 15 percent per hour, and from these figures it can be shown that the average concentration of caffeine in human beings is about $10^{-5}$ M, but it may go as high as $10^{-4}$M during part of the day for heavy coffee drinkers. In such individuals, caffeine is known to be present in gonads and fetuses.

Caffeine has been shown to have mutagenic effects (Andrew, 1959). It breaks human chromosomes in tissue culture, it breaks plant chromosomes (Kihlman, 1966), and it is a mutagen in bacteria (Novick, 1955) and *Drosophila* (Andrew, 1959), but not in mice (Cattanach, 1962). There exists a pronounced synergistic effect between caffeine and irradiation in bacteria (Shankel, 1962), and caffeine is known to inhibit DNA repair enzymes in bacteria (Shimada and Takagi, 1967). If mouse L cells are irradiated with ultraviolet (UV) light and plated on medium containing caffeine, cell killing is increased. Thus UV irradiation and caffeine have a synergistic effect on mammalian cells also (Rauth, 1970). On the other hand, caffeine does not seem to induce dominant lethals in mice (Rohrborn, 1968; Adler, 1970).

However, other scientists have found a positive effect in mice (Kuhlmann et al., 1968). Obviously, a weak mutagen can produce conflicting effects, and it is extremely difficult to prove its mutagenicity.

## Conclusion

There is an urgent need to organize mutagenesis data from current literature. This will make it possible to establish priorities as to which compounds in daily use and in the environment should be tested first. It will also help us to evaluate test systems. The test systems selected need to be scrutinized constantly for possible false results. New test systems should be developed continuously, and when these become routine, the compounds which gave negative results in the previous test systems should be reexamined.

Since point mutations are transmitted to the offspring with a higher probability than chromosome aberrations, the selected test systems should contain a method by which point mutations are detected. It is both costly and laborious to use mammals as a general screening system for the induction of point mutations. At present, it is only practical to use microorganisms in tests for the induction of point mutations. But in order to make the extrapolation from microorganisms to mammals, it is necessary to perform comparative mutagenesis studies with selected chemical mutagens in both types of organisms.

## References

Adler, I. D. (1970). In _Chemical Mutagenesis in Mammals and Man_. Berlin-Heidelberg-New York, Springer-Verlag.
Ames, B. N. (1971). In _Chemical Mutagens, Principles and Methods for Their Detection_. New York, Plenum Pub. Corp.
Andrew, L. E. (1959). _Amer. Natur._, _43_, 135.
Benzer, S. (1961). _Proc. Nat. Acad. Sci._, U.S., _47_, 403.
Brusick, D. (1971). EMS 2nd Ann. Mtg., Abstract No. 19.
Cattanach, B. M. (1959). _Z. Vererbungsl._, _90_, 1.
Cattanach, B. M. (1962). _Z. Vererbungsl._, _92_, 215.

Cumming, R. B., and M. F. Walton (1971). EMS 2nd Ann. Mtg., Abstract No. 2.

de Serres, F. J., and R. S. Osterbind (1962). Genetics, 47, 793.

de Serres, F. J. and H. V. Malling (1971). In Chemical Mutagens, Principles and Methods for Their Detection. New York, Plenum Pub. Corp.

Ehling, U. H., R. B. Cumming, and H. V. Malling (1968). Mutat. Res., 5, 417.

Fishbein, L., W. G. Flamm, and H. L. Falk (1970). Chemical Mutagens: Environmental Effects on Biological Systems. New York-London, Academic Press.

Freese, E. B. (1961). Proc. Nat. Acad. Sci., U.S., 47, 540.

Freese, E. B., and M. Cashel (1964). Biochim. Biophys. Acta, 91, 67.

Gabridge, M. G., and M. S. Legator (1969). Proc. Soc. Exp. Biol. Med., 130, 831.

Generoso, W. M., S. W. Huff, and S. K. Stout (1971). Mutat. Res., 11, 411.

Generoso, W. M., and W. L. Russell (1969). Mutat. Res., 8, 589.

Kihlman, B. A. (1966). Action of Chemicals on Dividing Cells. Englewood Cliffs, New Jersey, Prentice-Hall, Inc.

Kuhlmann, W., H. G. Fromme, E. M. Heege, and W. Ostertag (1968). Cancer Res., 28, 2375.

Lorkiewicz, A., and W. Szybalski (1961). J. Bacteriol., 82, 192.

Malling, H. V. (1966a). Mutat. Res., 3, 470.

Malling, H. V. (1966b). Mutat. Res., 3, 537.

Malling, H. V. (1967). Mutat. Res., 4, 265.

Malling, H. V., E. H. Y. Chu, and D. Wild (1971). EMS 2nd Ann. Mtg., Abstract No. 20.

Malling, H. V., and G. E. Cosgrove (1970). In Chemical Mutagenesis in Mammals and Man. Berlin-Heidelberg-New York, Springer-Verlag.

Malling, H. V., and F. J. de Serres (1968). Mutat. Res., 5, 359.

Malling, H. V., and F. J. de Serres (1969). Proc. Int. Bot. Congr., 11, 139.

Mrak, E. M. (1969). Report of the Secretary's Commission on Pesticides and Their Relationship to Environmental Health. Washington, D.C.: U. S. Department of Health, Education and Welfare.

Novick, A. (1956). Brookhaven Symp. Biol., 8, 201.

Rauth, A. M. (1970). In Current Topics in Radiation Research. Amsterdam, North Holland Pub. Co.

Rohrborn, G. (1968). Proc. Int. Congr. Genetics, Tokyo, 1, 103.
Russell, W. L. (1951). Cold Spring Harbor Symp. Quant. Biol., 16, 327.
Russell, W. L. (1967). Brookhaven Symp. Biol., 20, 179.
Russell, W. L., and R. B. Cumming (1971). EMS 2nd Ann. Mtg., Abstract No. 3.
Sega, G. A., and W. R. Lee (1970). Genetics, 64, 558.
Shankel, D. M. (1962). J. Bacteriol., 84, 410.
Shimada, K., and Y. Takagi (1967). Biochim. Biophys. Acta, 145, 763.
Somers, C. E., and T. C. Hsu (1962). Proc. Nat. Acad. Sci., U.S., 48, 937.
Webber, B. B., and F. J. de Serres (1965). Proc. Nat. Acad. Sci., U.S., 53, 430.

# THE DETECTION OF MUTATIONS WITH NON-MAMMALIAN SYSTEMS*

*Frederick J. de Serres*

Some of the advantages and disadvantages of various test systems for mutagenicity have already been alluded to (Malling, this volume). In general, non-mammalian systems have the advantages of speed, ease of testing, and generation of fairly precise results. Their disadvantage is a potential lack of relevance to man due to differences in physiology and metabolism. The following survey of non-mammalian test systems presents those that have been found to be most useful.

## Prokaryote Systems

### 1. DNA TRANSFORMATION

Genetic transformation in bacteria involves the transfer of genes from ruptured cells of donor bacteria to recipient bacteria which then develop the characteristics of the donor. Usually the following three species are used: Haemophilus influenzae, Diplococcus pneumoniae, and Bacillus subtilis. (For a review see Herriott, 1971.)

Mutation induction can be measured by testing for resistance to antibiotics (Hotchkiss, 1951; Hotchkiss and Evans, 1958; Hsu and Herriott, 1961; Ravin and Iyer, 1961) or for reversion of nutritional auxotrophs to wild type (Michalka and Goodgal, 1969; Anagnostopoulos and Crawford, 1961). This latter method has been used extensively to score mutation in a series of closely linked genes in B. subtilis where the operon for tryptophan synthesis is closely linked to the gene controling histidine synthesis. Muta-

*Sponsored by the U.S. Atomic Energy Commission under contract with the Union Carbide Corporation.

tions in three of the genes in the tryptophan operon will cause the cells to accumulate intermediates which fluoresce blue (Anagnostopoulos and Crawford, 1961; Carlton, 1967).

The test for mutation induction is simple and straightforward. The DNA is isolated from wild type cells and treated with a mutagen. Treated DNA is mixed in transformation medium with recipient cells which are then plated out on supplemented medium. The plates are incubated and screened for the presence of fluorescent colonies with a UV mineral light.

Reversion of nutritional mutants can also be scored by plating cells on a selective medium on which only revertants can grow. The mutagen is added to a sterile paper disk placed in the center of the plate. A concentration gradient develops during incubation, and a ring of revertant colonies will appear outside a ring of complete inhibition. The number of colonies can be easily counted (see Freese and Strack, 1962).

## 2. BACTERIOPHAGE

These are hi hly specific viruses which infect specific types of bacteria. They are very useful for mutation studies because they are easy to grow, have short generation times, and very large populations can be studied.

A particularly useful system for studying mutation induction is the assay for plaque mutants after infection of *Escherichia coli* with bacteriophage $T_4$. The wild type phage produces plaques a few millimeters in diameter with fuzzy edges, and certain mutants (*r*) produce much larger plaques with sharp edges (Benzer, 1955, 1959, 1961). Mutants at the *rII* locus comprise about 70 percent of the spontaneous *r* mutants and occupy two adjacent cistrons. The inability of *rII* mutants to grow in bacteria containing the genome of lambda phage provides a screening system for the analysis of reverse mutation. A tester set of mutants which revert by various mechanisms can be selected for use in spot tests on petri plates. If the compound being tested is mutagenic, a halo of revertant plaques will form around the spot containing the mutagen. If the chemical is specific and produces only a particular type of genetic alteration, halos will be found in the plates containing some of the tester strains and not others. Such a procedure provides a rapid and efficient screening system to detect the mutagenic activity of unknown agents.

## 3. BACTERIA

The most widely used bacterial systems in studies of mutagenesis are reverse mutation systems. One of the best systems for such studies involves the use of histidine-requiring mutants in the C gene of the histidine operon of *Salmonella typhimurium*. Alterations in aminotransferase (C gene) mutants in the histidine operon have been identified on the basis of specific revertibility tests after treatment with chemical mutagens (Whitfield et al., 1966; Ames and Whitfield, 1966). Nonsense, missense, and frameshift mutants are known. Each mutant reverts as a result of a particular type of genetic alteration. By selection of mutants that revert at high frequencies after treatment, a tester set can be developed which is capable of detecting and identifying the mutagenic activity of both strong and weak mutagens. (For more extensive discussion of the use of bacterial systems for mutation detection see Ames, this volume.)

A suspension of each mutant in the tester set is spread over the surfaces of petri plates containing minimal medium supplemented with a trace of histidine (0.2 micromole per ml) so that the background lawn can grow slightly and any inhibition by the compounds to be tested can be seen. The chemicals to be tested are dissolved in distilled water, and a drop of each is placed at spaced intervals around the periphery of each plate. After two days incubation at 37° C, mutagenic activity is indicated by the presence of a ring of revertant colonies around the spot where a chemical was added. Weak mutagens will produce only a few revertants, and strong mutagens will produce many revertants. The type of genetic alteration produced by each chemical is indicated by the particular mutants in the tester set which are induced to revert. This test system is well adapted for routine screening of chemicals for mutagenic activity. The major disadvantage is that it is a reverse mutation system which can detect only particular types of genetic alterations.

## Eukaryote Systems

### 1. ASCOMYCETES

One advantage of these nonbacterial but microbial systems is the fact that their genetic information is organized into a discrete nucleus and, insofar as is known, have chromosomes whose structure and morphology are similar to those of higher organisms. Some have life cycles that alternate between a haploid genome (as in human gametes) and a diploid genome (as in human somatic cells). Some haploid strains can be made to fuse to produce heterokaryons with two genetically different nuclei in the same cytoplasm rather than in the same nucleus. These heterokaryons lend themselves readily to certain types of genetic tests.

A variety of assay systems have been developed for studies on mutagenesis; the following have been particularly useful.

#### a. Early Genes in the Purine Biosynthetic Pathway in Saccharomyces cerevisiae

In yeast there are two recessive genes, ad-1 and ad-2, which cause a requirement for adenine and which result in the accumulation of a red pigment. When a haploid red clone is plated on YEP medium (1 percent yeast extract, 2 percent bactopeptone, 8 percent glucose, and 1.5 percent agar), some of the colonies are white or pale pink, and these retain their phenotype on transfer and plating (Roman, 1956). In addition, red colonies growing on solid medium form white and pale pink papillae from which stable mutants can also be obtained.

Genetic analysis has shown that some of the white or pale pink variants are due to reversion of the ad-1 or ad-2 allele to adenine independence. But when a stable ad-1 or ad-2 allele was used, all of the white or pale pink colonies were found to result from mutation in one of five genes controling earlier steps in purine biosynthesis: ad-3, ad-4, ad-5, ad-6, or ad-7. The mutant genes in the pale pink colonies were found to be intermediate alleles at these same loci.

To screen for mutagenicity, suspensions of cells of a stable ad-1 or ad-2 mutant are treated for the same period of time with varying concentrations of the chemical. The treated cells and untreated controls are plated on the surface of YEP agar and after incubation the plates are screened for the presence of white and pale pink colonies among the red colonies. The presence of a higher frequency of such variants than is present in the plates of the untreated control indicates mutagenic activity. The genotypes of the white and pale pink variants can be determined in crosses with the appropriate tester strains to determine the spectrum of induced genetic alterations.

This is a forward mutation system and presumably is capable of detecting any type of genetic alteration which will produce gene mutation by intragenic alteration.

b. Tester Set of ad-3B Mutants in Neurospora crassa

The genetic alterations in a series of nitrous acid induced ad-3B mutants of *Neurospora crassa* were identified on the basis of specific revertibility after treatment with chemical mutagens (Malling, 1966; Malling and de Serres, 1968). A tester set has been developed consisting of eight mutants which revert at high frequency after treatment. Four mutants revert only by base pair substitutions: two by AT→GC and the other two by GC→AT. Two mutants revert only by base pair insertion or deletion, and two mutants revert only spontaneously.

The tester set is used to test for mutagenic activity and for specificity of this activity. Suspensions of each mutant strain are treated for the same period of time with varying concentrations of the chemical. The treated suspensions and untreated controls are plated on minimal medium supplemented with a trace of adenine (2.0 mg/l) to permit a little background growth. After 7 to 9 days incubation at 30° C, the number of revertant colonies are counted. A higher frequency of revertant colonies in the treated series than in the untreated controls indicates mutagenic activity. Specificity is indicated when only those strains requiring the same type of genetic alteration to revert are affected. Strong mutagens will produce a high frequency of reversion; weak mutagens, a low frequency.

This method is useful for rapid screening tests for mutagenicity. However, because it screens for particular types of genetic alterations, a negative test does not mean that the compound is not mutagenic. It is possible for a chemical to produce genetic alterations which would not be detected by any of the eight strains in the tester set.

### c. Resistance to 4-Methyl Tryptophan in *Neurospora crassa*

*Neurospora* has at least two systems which enable it to accumulate amino acids and to concentrate them up to 100 times the level present in the surrounding medium. The *mtr+* locus controls the transport system which preferentially takes up aromatic amino acids (Lester, 1966; Stadler, 1966). However, other amino acids can be concentrated by this system as well as certain amino acid analogs. As these amino acid analogs are concentrated by the cell, synthesis of the normal amino acid is stopped as growth is inhibited. Strains resistant to the effect of the amino acid analog 4-methyl-tryptophan (4MT) result from mutation at the *mtr* locus (*mtr+* → *mtr*; the *mtr* mutants are recessive to the wild type allele.

To use this system to determine the mutagenicity of chemicals, treated conidia and untreated controls are plated in the presence of 4MT. At high concentrations of this analog, growth of *mtr+* conidia is greatly inhibited and the only colonies formed are those resulting from mutations to resistance (*mtr*). Mutagenicity is indicated by a higher frequency of colonies (*mtr* mutants) in the treated series than in the control series.

This is a forward mutation system which is capable of detecting any type of genetic alteration resulting in the inactivation (partial or complete) of the mtr+ allele by intragenic alteration. By using a heterokaryon heterozygous at the mtr locus, the system can be made to detect extragenic mutations where the mtr+ locus is inactivated by chromosome deletion. Thus the production of recessive lethal mutations at this locus, both by gene mutation and by chromosome breakage and deletion, can be detected. This is a simple test system which can be used for routine screening and which is capable of detecting a broad spectrum of genetic damage. There is not yet any simple way to distinguish the recessive lethal mutations resulting from gene mutations from those resulting from chromosome deletion. Thus, although a heterokaryon heterozygous for mtr is theoretically capable of detecting all types of genetic damage, no characterization of the mtr mutants is possible.

d. 2-Thioxanthine Resistance in *Aspergillus nidulans*

In *Aspergillus nidulans*, 2-thioxanthine produces an effect on conidial pigmentation. Green-conidiating wild type strains conidiate yellow in the presence of 2-thioxanthine, but certain mutant strains which are resistant to 2-thioxanthine conidiate green. Resistance results from mutation at one of eight gene loci (Alderson and Scazzocchio, 1967). The resistant mutants have been phenotypically characterized as lacking the enzyme xanthine dehydrogenase (XDH) and being unable to grow on hypoxanthine as a sole source of nitrogen. Failure to grow on certain other sole nitrogen sources divides the XDH mutants into three classes: (1) hx mutants which fail to grow on hypoxanthine (two genes), (2) uaY mutants which also lack urate oxidase activity and fail to grow on hypoxanthine or uric acid (one gene), and (3) cnx mutants which also lack nitrate reductase activity and fail to grow on hypoxanthine or nitrate (five genes).

To use this test system to assay for mutagenicity, conidia from a haploid wild type are treated with different concentrations of the chemical for the same time period, and treated conidia and untreated controls are plated on media containing

2-thioxanthine to give about 200 colonies per plate. After three or four days incubation at 37°C, the platings are examined for colonies which are resistant (conidiated green); these resistant colonies are picked and plated to establish pure strains. Conidia from the mutant strains are tested for growth on five different supplemented media to determine their growth requirements. In this way each mutant is assigned to one of eight different loci.

e. Recessive Lethal Mutations over the Entire Genome in *Neurospora crassa*

By using a two-component heterokaryon of *Neurospora* with each component marked with appropriate biochemical and morphological markers, it is possible to determine the frequency of recessive lethal mutations occurring over the entire genome (Atwood, 1949). The heterokaryon has an arginine (*arg-6*) requirement in component I, and a methionine (*meth-7*) requirement and a morphological marker amycelial (*amyc*) in component II. Recessive lethal mutations induced in component II are propagated indefinitely because of the presence of the normal alleles in component I. A two-component heterokaryon of this type produces three types of colonies: one heterokaryotic type containing at least one nucleus of each genotype, and two homokaryotic types containing only nuclei of component I genotype or component II genotype. By plating on minimal medium, only the heterokaryotic conidia will grow; by plating on minimal medium supplemented with arginine or methionine the homokaryotic conidia will grow. If no recessive lethal mutation is present in the *meth amyc* nucleus, this component forms a tiny morphologically distinct colony on minimal media containing methionine. If a recessive lethal mutation has been induced, only morphologically normal (large) colonies are found.

The frequency of heterokaryotic colonies which do not produce the tiny *amyc* colonies on plating gives a minimal estimate of the recessive lethal frequency, since lethal mutations occurring in treated conidia with more than one *meth amyc* nucleus are not detected. However, since the average number of nuclei per conidium is about 2.3, it seems likely that the majority of the recessive lethal mutations are detected.

To test for mutagenicity with this test system, conidia of the heterokaryon are treated and plated on minimal medium along with untreated controls. A random sample of about 100 to 150 colonies are isolated and subcultured. Conidia from each isolate are then plated to screen for the presence of amyc colonies on minimal medium supplemented with methionine. Plates containing normal but no amycelial colonies are scored as lethals. Mutagenicity is indicated by a higher frequency of lethals in the treated series than in the untreated controls.

This is a forward mutation system capable, in theory, of detecting any type of genetic alteration which results in the inactivation of any of the genes in the genome. Homology tests of recessive lethal mutations in general show that they consist of both gene mutations and chromosome deletions that may cover more than one locus (Atwood and Mukai, 1953). This test system provides a simple assay for mutagenicity which tests for an effect on a very large number of genes. It is not possible to further characterize the recessive lethal mutants obtained.

### f. ad-3 Mutants of Neurospora crassa

Mutation of two genes in the purine biosynthetic pathway, ad-3A and ad-3B, results not only in a requirement for adenine but in the accumulation of a reddish purple pigment in the vacuoles of the mycelium. Because the ad-3 mutants are phenotypically distinguishable from wild type on the basis of this pigment accumulation, it has been possible to develop a direct method for their recovery (de Serres and Kolmark, 1958). The spectrum of ad-3 mutants recovered with this method ranges from nonleaky mutants having a complete requirement for adenine to leaky mutants with only partial requirements. Mutants at the ad-3B locus show allelic complementation and have a linear complementation map with 17 complons (de Serres et al., 1967). A correlation has been found between complementation pattern and genetic alteration at the molecular level (Malling and de Serres, 1967, 1968), so that the heterokaryon test for allelic complementation can be used to obtain presumptive evidence of the spectrum of genetic alterations at the molecular level.

By using a two-component heterokaryon heterozygous at both the ad-3A and ad-3B loci, it is possible to obtain not only those recessive mutations resulting from intragenic alteration but also those resulting from chromosome deletion (de Serres and Osterbind, 1962). Thus, with this test system, it is possible to study forward mutation resulting from a wide variety of genetic alterations.

To assay for mutagenicity, the conidia from the two-component heterokaryon are treated with varying concentrations of a chemical for a constant period of time. The treated conidia and untreated controls are incubated separately in 10 l liquid medium (made viscous with 0.15 percent agar) in 12 l Florence flasks for 7 days in the dark at 30°C. The contents of each jug are analyzed for the presence of ad-3 mutants by pouring 1500 ml aliquots into large white photographic developing trays and scanning for the presence of small reddish purple colonies among the white background colonies; 10 ml samples of the background colonies are reserved for direct counts to estimate the total number of colonies per flask. The purple colonies are subcultured and plated out to make them homokaryotic for the adenine requirement, and a subculture of this homokaryotic derivative is put into stock culture. A series of genetic tests is then performed to determine genotype and allelic complementation and to distinguish the point mutations from the chromosome deletions.

The ad-3 test system undoubtedly provides the most sophisticated test for mutagenicity. The ability to obtain precise quantitative data makes it possible to obtain dose-response curves not only for the induction of ad-3 mutants in general, but also for each of the subclasses of point mutations and chromosome deletions. In addition it is possible to obtain presumptive evidence of the spectrum of genetic alterations, resulting in point mutation at the molecular level.

This test system provides a genetic analysis "in depth" of the spectrum of mutation resulting in the production of recessive lethal mutations at two specific loci in the genome of Neurospora.

g. Red Adenine Mutants of Yeast

Forward mutation in genes controling the biosynthesis of adenine which results in the formation of red-pigmented colonies has been studied extensively in yeast. In Saccharomyces cerevisiae the two loci are ade1 and ade2, and in Schizosaccharomyces pombe the two loci are ade6 and ade7. In both yeasts the two loci are unlinked, in contrast to the analogous purple-adenine mutants ad-3A and ad-3B in Neurospora which are closely linked. To study mutation at the red-adenine loci in yeast, mutagen-treated haploid cells are plated onto complete medium to yield a total of about 500 colonies per plate. Mutant colonies or colony sectors can be readily identified, since wild type cells form white to cream-colored colonies. Frequencies of spontaneous mutations have been estimated to be less than $10^{-7}$ (Loprieno et al., 1969b), whereas the frequencies of chemically-induced mutation at these two loci are in the range $10^{-4}$ to $10^{-2}$ (Loprieno et al., 1969a). The only problem with the assay system is the number of colonies that can be scored conveniently to detect mutation induction after mutagenic treatment: with potent mutagens, mutants can be detected in as few as 10 to 20 plates; to detect mutagenic activity of a weak mutagen that gives only a 10-fold increase over the spontaneous frequency ($10^{-6}$) would require screening 2000 plates to detect a single mutant! The simplicity of the assay, however, for detection of potent mutagenic activity would make this test system an especially useful primary screen with a eukaryotic organism.

2. DROSOPHILA

Some of the most sophisticated assay systems for studying mutagenesis have been developed with the fruit fly Drosophila melanogaster. Methods have been developed for detecting lethal and visible gene mutations, chromosomal rearrangements, and chromosome loss. (For a review, see Abrahamson and Lewis, 1971.)

One of the most useful and meaningful assays involves the screening for sex-linked recessive lethals. This test has the advantages of yielding results in a few generations, testing a large number of loci, and objectivity. Treated males are mated to females homozygous for a balancer chromosome. The $F_1$ females and males are mated with each other, and the $F_2$ progeny are examined with a hand lens or stereoscopic microscope. The number of cultures which lack normal-eyed (non-Bar) males provide an estimate of the number of treated X-chromosomes which carry one or more X-linked recessive lethal mutations. Similar tests have been devised to screen for recessive lethal mutations on the autosomes (Abrahamson and Lewis, 1971).

Reciprocal translocations in *Drosophila* can be detected in the second generation after treatment by a standard method utilizing recessive marker genes in the autosomes. By using the eye color mutants brown (bw) and scarlet (st), which are located in the second and third chromosomes, it is possible to detect not only translocations between the two major autosomes but also between the Y chromosome. The advantage of the *Drosophila* assay is that cultures scored as translocations can be tested further to verify the rearrangement by cytological analysis of the giant salivary gland chromosomes.

### 3. HABROBRACON

The parasitic wasp *Habrobracon* is another organism that is extremely useful for mutagenicity studies. Studies with *Habrobracon serinopae* show that it is superior to *H. juglandis* used in early work (Smith and von Borstel, 1971). *H. serinopae* is relatively disease free and has a short life cycle (7-1/2 days at 28° C); the females live for about two months and lay about twice as many eggs per day as *H. juglandis*. *Habrobracon* can be used to study the same types of events as in *Drosophila*: dominant lethality, recessive lethal mutations, visible mutations, and chromosome rearrangements. In *Habrobracon*, unmated females produce haploid male offspring. When treated females are set unmated for oviposition, hatchability is a measure of total lethality (dominant and recessive lethal mutations). When treated females are set for oviposition after being mated to untreated males, hatchability is a measure of dominant lethality only.

A quick analysis of mutagenic action can be made by comparing hatchabilities between fertilized and unfertilized eggs after the eggs are immersed into a solution of a potential mutagen (Clark and Beiser, 1955). In this test, nuclear inactivation can be distinguished from cytoplasmic inactivation: nuclear inactivation will have a greater effect on the diploid egg nucleus than on the haploid egg nucleus, whereas cytoplasmic inactivation affects both types of egg nuclei equally.

4. TRADESCANTIA

*Tradescantia* plants heterozygous for flower color provide a useful test system to detect mutagenic activity. Young flower buds on intact plants or on cuttings can be exposed to various mutagens in either a gaseous or aqueous state. *Tradescantia* is relatively easy to grow under a wide range of environmental conditions, blooms continuously throughout the year, has twelve large chromosomes and a cellular radiosensitivity similar to that of mammalian cells. A special strain heterozygous for flower color can be used for easy detection of somatic mutations in both petal and stamen hair tissues using only a dissecting microscope and simple laboratory techniques (Ichikawa et al., 1969; Mericle and Mericle, 1967; Nayar and Sparrow, 1967; Sparrow et al., 1968). After exposure of the young flower buds to mutagenic treatment, somatic mutations and morphological change in petals and stamen hairs may be scored throughout a 10- to 20-day posttreatment period, making it possible to assay for genetic damage induced during various meiotic and mitotic stages during microspore development.

## References

Abrahamson, S., and E. B. Lewis (1971). In Chemical Mutagens, Principles and Methods for Their Detection, A. Hollaender, editor. New York, Plenum Pub. Corp.

Anagnostopoulos, C., and I. P. Crawford (1961). Proc. Natl. Acad. Sci., U. S., 47, 378.

Alderson, T., and C. Scazzocchio (1967). Mutation Res., 4, 567.

Ames, B. N., and H. J. Whitfield, Jr. (1966). Cold Spring Harbor Symposium Quant. Biol., 31, 221.

Atwood, K. C. (1949). Biol. Bull., 97, 254.

Atwood, K. C. and F. Mukai (1953). Proc. Natl. Acad. Sci., U. S., 39, 1027.

Benzer, S. (1955). Proc. Natl. Acad. Sci., U. S., 41, 344.

Benzer, S. (1959). Proc. Natl. Acad. Sci., U. S., 45, 1607.

Benzer, S. (1961). Proc. Natl. Acad. Sci., U. S., 47, 403.

Carlton, B. (1967). J. Bacteriol., 94, 660.

Clark, A. M., and W. C. Beiser, Jr. (1955). Science, 121, 469.

de Serres, F. J. and H. G. Kolmark (1958). Nature, 182, 1249.

de Serres, F. J. and R. S. Osterbind (1962). Genetics, 47, 793.

de Serres, F. J., H. E. Brockman, W. E. Barnett, and H. G. Kolmark (1967). Mutation Res., 4, 415.

Freese, E., and H. B. Strack (1962). Proc. Natl. Acad. Sci., U. S., 48, 1976.

Herriott, R. M. (1971). In Chemical Mutagens, Principles and Methods for Their Detection, A. Hollaender, editor. New York, Plenum Pub. Corp.

Hotchkiss, R. D. (1951). Cold Spring Harbor Symposium Quant. Biol. 16, 457.

Hotchkiss, R. D., and A. H. Evans (1958). Cold Spring Harbor Symposium Quant. Biol., 23, 85.

Hsu, Y. C. and R. M. Herriott (1961). J. Gen. Physiol., 45, 197.

Ichikawa, S., A. H. Sparrow, and K. H. Thompson (1969). Radiation Botany, 9, 195.

Lester, G. (1966). J. Bacteriol., 91, 677.

Loprieno, N., R. Guglielminetti, S. Bonatti, and A. Abbondandolo (1969a). Mutation Res., 8, 65.

Loprieno, N., S. Bonatti, A. Abbondandolo, and R.

Guglielminetti (1969b). Molec. Gen. Genetics, 104, 40.
Malling, H. V. (1966). Mutat. Res., 3, 470.
Malling, H. V. and F. J. de Serres (1967). Mutation Res., 4, 425.
Malling, H. V. and F. J. de Serres (1968). Mutation Res., 5, 359.
Mericle, L. W. and R. P. Mericle (1967). Radiation Botany, 7, 449.
Michalka, J. and S. H. Goodgal (1969). J. Mol. Biol., 45, 407.
Nayar, G. G. and A. H. Sparrow (1967). Radiation Botany, 7, 257.
Ravin, A. W. and V. N. Iyer (1961). J. Gen. Microbiol., 26, 277.
Roman, H. (1956). Compt. Rend. Trav. Lab. Carlsberg Ser. Physiol., 26, 299.
Smith, R. H., and R. C. von Borstel (1971). In Chemical Mutagens, Principles and Methods for Their Detection, A. Hollaender, editor. New York, Plenum Pub. Corp.
Sparrow, A. H., K. P. Baetcke, D. L. Shaver, and V. Pond (1968). Genetics, 59, 65.
Stadler, D. (1966). Genetics, 54, 677.
Whitfield, H. J., Jr., R. G. Martin, and B. N. Ames (1966). J. Mol. Biol., 21, 335.

# A BACTERIAL SYSTEM FOR DETECTING MUTAGENS AND CARCINOGENS*

*Bruce N. Ames*

It is generally agreed that it is extremely important to minimize exposure of the human gene pool to mutagens. The problems of testing mutagens are in many ways fundamentally different from those of the usual toxicological and pharmacological tests, in part because mutagenesis must be monitored in the descendents of large populations. Probably no one test system will ever be completely satisfactory, and each of the several systems that will be used has some theoretical and practical advantages.

Test systems for mutagens using microorganisms or phage can be very sophisticated, extremely sensitive, simple, and economical (a variety of methods are described in Hollaender, 1971). We have developed a bacterial test system for mutagens that has proven to be very effective in detecting known mutagens and carcinogens (Ames, 1971). We will discuss the advantages and limitations of the test, the mutagens detected, and in particular the theoretical reasons for believing that the microorganisms are a useful system for detecting potential mutagens and carcinogens for humans.

In ordinary toxicology or pharmacology, it is desirable to use a mammal closely related to humans, as protein and metabolic interactions differ from one animal to the next. Usually the target protein of the compound being tested is not known. When testing for mutagens, on the other hand, the target is known: mutagens react with DNA, and the DNA has the same four bases and the same Watson-Crick structure in all living things. Thus it is quite reasonable to use bacteria (or any living organism) for the detection

*This work was supported by AEC grant AT(04-3).

of potential mutagens for humans, with certain obvious limitations that will be discussed below.

As almost all of the numerous mutagens that can be detected in the bacterial system are carcinogens, we believe that the bacteria will be extremely useful for predicting potential carcinogens for humans, with again some limitations that will be discussed below.

### Advantages of Using a Particular Set of Four Strains We Have Developed in *Salmonella typhimurium*

#### 1. SIMPLICITY: THE PLATE TEST

A small sample (about 1 mg) of the suspected mutagen can be placed in the center of a petri plate that has been seeded with a lawn of bacteria that cannot grow because of a mutation. If the mutagen can cause that particular mutation to revert in an occasional bacterium, it will enable that bacterium to grow and form a colony. Therefore a circle of colonies will appear around the spot of mutagen after about a day and a half of incubation. The diffusion of the mutagen in the agar allows one, in effect, to test a wide range of concentrations on a single plate. Compounds that are insoluble in water can be dissolved in dimethylsulfoxide, which is not very toxic for the bacteria, and up to 0.5 ml added to the plate.

#### 2. COMPREHENSIVENESS: TESTER STRAINS FOR A VARIETY OF TYPES OF DNA ALTERATIONS

We have screened hundreds of different histidine-requiring mutants and have picked a set of four (Ames, 1971) that are particularly suitable in plate tests (low spontaneous reversion rate, high sensitivity, etc.) for detecting mutagens which cause specific changes: base pair substitutions (TA1530 = hisG46 uvrB) (e.g., streptozotocin and other alkylating agents, sodium nitrite) or insertion or deletion of one or two base pairs (TA1531 = hisC207 uvrB, TA1532 = hisC3076 uvrB, TA1534 = hisD3052 uvrB) (atabrine, nitroquinoline-N-oxide*, hycanthone*, nitroso-fluorene, benz(a)anthracene-5,6-epoxide, and other polycyclic carcinogens).

---

*See footnote on page 59.

We have also developed a test for large deletions (no mutagens detected solely by this test have been found as yet), and we are perfecting a test for gene duplication. In addition a general test for forward mutagenesis has been developed (resistance to azetidine-carboxylic acid or to hydrazino-imidazole-propionic acid) that picks up all of these types (Ames, 1971, unpublished). It is based on the destruction of a permease gene. We hope that we can design bacterial tests that pick up all classes of proximal mutagens.

The four tester strains recommended now (and the comparison described in Section A4) add comprehensiveness to the sensitivity and simplicity of a specific back mutation test. A comprehensive set of back mutation tests is superior to a forward mutation test because of the simplicity and the greatly increased sensitivity due to a lower spontaneous background (see also effect of repair). In addition, many forward mutation test systems for mutagens (e.g., those detecting dominant mutations) miss particular classes of mutagens. For example, a mutation being dominant in bacteria is usually due to a protein with a changed specificity rather than a protein that has been made non-functional. Agents that cause frameshift mutations (many of the proximal polycyclic hydrocarbon carcinogens fall into this category) do not show up on such a test because they cause a non-functional protein. One cannot obtain streptomycin-resistance (a well-known dominant mutation) in bacteria by using a frameshift mutagen (Silengo et al., 1967). Thus it is important to know that any particular test will in fact detect the variety of known mutagens.

3. SENSITIVITY: STRAINS LACKING REPAIR

A very large population is essential in testing a mutagen to obtain adequate sensitivity. If one specific gene, out of 100,000 genes, is being monitored for mutation in an animal, then at the very high level of mutation where every organism has one

---

*Nitroquinoline-N-oxide and hycanthone were detected by Hartman et al. (1971) with hisD3052 and we have added a uvrB mutation to this strain.

gene damaged, only 1 in 100,000 animals will have a damage in the particular gene monitored. One can obtain this sensitivity in bacteria or in bacterial viruses or in tissue culture, but not easily in animals. In the bacterial test system, even rare mutational events may be detected readily, since about $5 \times 10^8$ bacteria are exposed to the mutagen on a petri plate. Even if only a few bacteria are mutated at the appropriate nucleotide (out of about $4 \times 10^6$ base pairs), each gives rise to a colony that can be observed in two days. We have developed a method of increasing this sensitivity another one or two orders of magnitude. In bacteria (as in higher organisms) almost all of the primary damage to the DNA caused by a mutagen is repaired by the excision and recombination repair systems so that only a small percentage of the potential mutations are detected. We have introduced into our four bacterial tester strains, a defective excision repair system (*uvrB* gene deletion), so that they are 10 to 100 times more sensitive to mutagens than the parent strains.

## 4. THE COMPARISON OF STRAINS WITH AND WITHOUT EXCISION REPAIR

In addition to its great utility as a tester strain, the double mutant lacking repair (e.g., TA1534 = *hisD3052 uvrB*) can be compared with the histidine-requiring mutant with repair (e.g., *hisD3052*). This comparison is very useful in three different ways.

a. A difference in the zone of inhibition shown by a compound on the strains with and without the repair system furnishes evidence that the lethality of the compound is caused by reaction with DNA, independently of its action as a mutagen (e.g., see beta-propiolactone in Ames, 1971). We routinely monitor this, as compounds are known that react with the bacterial DNA yet are not mutagenic for the bacteria. We use TA1534 and *hisD3052* for this comparison, as TA1534 has the normal lipopolysaccharide (TA1530, TA1531, and TA1532 are lacking the *gal* operon). (See also Slater et al., 1971.)

b. The intercalating agents which do not react with DNA are equally effective on the strains with and without the repair system. On the other hand, the compounds such as the acridine-half-mustards that alkylate as well as intercalate are much more effective on the strains lacking repair. We have found that the carcinogens (and frameshift mutagens) 2-nitroso-fluorene and benzanthracene epoxide are much more effective as mutagens on the strains lacking repair, and we conclude from this that they are reacting with DNA, as well as intercalating.

c. The range of defects repaired by the repair system can be determined (Ames, 1971). This is useful also in investigating agents, such as caffeine, that inhibit repair. For example, 1 mg of nitrogen mustard gives 75 times as many histidine revertants of the double mutant lacking repair, hisG46 uvrB, as it does with the hisG46 strain. Nitrogen mustard when added to hisG46 in the presence of caffeine causes almost as pronounced an increase. Caffeine, though it potentiates mutagenesis by inhibiting repair, is not itself a mutagen. Methylation is not repaired by this repair system, though ethylation is to a small extent (Ames, 1971), since no increase is seen with TA1530 as compared to hisG46 (or with hisG46 with and without caffeine) when methylating agents are used (G. Ficsor, E. Zeiger, personal communications; B. Ames, unpublished). Thus in testing the effect of caffeine on repair in mammals, we suggest it is important to test it with a mutagen such as nitrogen mustard in addition to the reported tests with a methylating agent.

### Frameshift Mutagens and Their Relation to Polycyclic Hydrocarbons

Three of our tester strains are specific for various types of frameshift mutagens. We suspect frameshift mutagens will be an extremely important class of environmental mutagens. They are often non-reactive and relatively non-toxic in the usual sense, so they may have been overlooked in ordinary testing.

An addition or deletion of a base pair from the DNA can occur when there is a mispairing in a string of nucleotides during DNA replication. This results in a frameshift mutation. This process is enormously increased by the addition of an acridine type compound that intercalates in the DNA base pair stack and stabilizes mispairing. We have found that when an intercalating agent (an acridine) has also a side chain (a nitrogen half-mustard) that can react with DNA, it is a more potent mutagen for causing frameshift by one to two orders of magnitude (Ames and Whitfield, 1966). Polycyclic hydrocarbons are known to intercalate in DNA (Craig and Isenberg, 1970). It is also suspected that polycyclic hydrocarbons are metabolized in mammals to the true or proximal carcinogens that can react with DNA. We have demonstrated that 2-amino-fluorene, a well known carcinogen, is a frameshift mutagen in our bacteria, and that nitroso-fluorene, a known metabolite of this in the rat (and a more effective carcinogen, Miller and Miller, 1969) is hundreds of times more effective as a frameshift mutagen for bacteria (Ames and Miller, in preparation). The nitroso group is a reactive group and we believe that nitroso-fluorene is reacting with the DNA and intercalating by analogy with the acridine half-mustards. Epoxides of polycyclic hydrocarbons are active as frameshift mutagens (Ames, Sims, and Grover, submitted for publication.) We suspect that the proximal carcinogens derived from polycyclic hydrocarbons will be frameshift mutagens, and that they can be detected with our bacteria with great sensitivity.

There is a high degree of specificity in frameshift mutagens as to GC or AT strings, or addition or deletion of bases. The three tester strains (TA1531, TA1532, and TA1534) will detect with great sensitivity the variety of types of frameshift mutagens. The acridine half-mustards react with guanine and are detected with TA1531 and TA1532, while 2-nitroso-fluorene is completely ineffective against these strains but is extremely effective in reverting TA1534. Other compounds are effective in reverting TA1531 (which we think has a deleted GC pair), but not TA1532 (which appears to have an added GC pair), and vice versa.

## Agents That Have Been Shown to Be Mutagenic Using These Strains

Radiation: ultraviolet+, fast neutrons+, X-rays+
Methylating agents: N-methyl-N'-nitro-N-nitroso-guanidine+, streptozotocin+, methylmethane sulfonate+, N-nitroso-N-methyl-urethane+, methylazoxymethanol+

Ethylating agents: diethylsulfate+, ethylmethanesulfonate+
Other alkylating agents: β-propiolactone+, β-butyrolactone+, 1,3-propane sultone+, nitrogen mustard+, chloroethylamine+, dibromoethane (ethylenedibromide), captan*, ethyleneimine+
Non-alkylating agents: hydrazine+, hydroxylamine, nitrous acid
Intercalating agents: quinacrine (atabrine), 9-aminoacridine, hycanthone, 2-nitroso-fluorene+, nitroso-carbazole, benz(a)anthracene-5,6 epoxide+, nitroquinoline-N-oxide+, N-nitroso-diphenylamine, p-nitroso-diphenylamine, and a variety of acridines, fluorenes, benzacridines and anthracenes
Base analogs: 2-aminopurine, 5-bromouracil

### Reasons Why a Compound That Is Mutagenic or Carcinogenic in Humans May Be Missed in a Bacterial Screen

1. The compound tested is converted by metabolism in humans, but not in bacteria, to the true or proximal mutagen or carcinogen. This is an important point and many compounds will be missed in the direct bacterial test. If the metabolites of a compound are known, they also can be tested directly (e.g., 2-nitroso-fluorene is much more mutagenic than acetyl-amino-fluorene or 2-amino-fluorene (Ames, 1971; Ames and Miller, in preparation); benz(a)anthracene-5,6-epoxide is a good mutagen while benz(a)anthracene is not (Ames, Sims, and Grover, submitted for publication). Many proximal mutagens (e.g., cycasin, dimethylnitrosamine) can be detected by using these bacterial strains in the host mediated assay (Legator and Malling, 1971) where they are exposed to the metabolites of the compound _in vivo_. In addition, the proximal mutagen (e.g., dimethylnitrosamine) may be detectable by mixing a liver microsomal system (Malling, 1971) or tissue homogenates (Ficsor and Muthiani, 1971) with the bacterial tester strains. In addition, mixtures of metabolites or enzymatic reaction products can be

---

+ indicates agents known to be carcinogenic
*G. Ficsor, personal communication

tested on the bacteria for a rapid bioassay of the proximal mutagen (carcinogen).

2. The compound is taken up by humans and not by bacteria. We suspect differential uptake will not be of very great importance. All cells have very selective active transport systems designed to take up normal metabolites past the cell membrane. The usual mutagen is not likely to enter the cell through these systems, in either humans or bacteria. We think that, in order for a compound to be mutagenic in any system involving whole cells, the mutagen must dissolve directly in the cell membrane and thus circumvent the specific transport. Mutagens that have been identified are generally very soluble in organic solvents and would presumably dissolve in all cell membranes (see list in Section C, for example). In a eucaryotic cell, of course, there would be the cellular membrane, cytoplasm, and the nuclear membrane to pass. There is reason to believe that both bacterial and eucaryotic DNA are associated with membrane during replication.

3. The compound causes chromosome abnormalities (breakage or translocation or duplication) and this would be unique to eucaryotic cells. It remains to be seen what mutagens fall into this class and do not cause point mutations or deletions or react with the DNA in bacteria.

4. The compound is detoxified by bacteria but not by humans. This is possible, but seems relatively unlikely (see E3).

**Reasons That a Compound That Is Mutagenic for Bacteria May Not Be Mutagenic for Humans**

1. The compound is detoxified in some way before it reaches the germ cells. This certainly should be important. Nevertheless, many such compounds may still be carcinogenic; in fact, some compounds may be mutagenic for somatic cells and cause cancer but not reach the germ cells and thus not be mutagenic. It is clear that most of the compounds that are mutagenic in the bacterial test are carcinogenic (see Section C and the excellent review by Miller and Miller, 1971). It remains to be

seen how many mutagens detected by the bacterial screen will not produce cancer in animals.

2. The bacteria take up the compound, but human cells do not. We don't think this will be too important, though it is certainly a possibility (see discussion in D2).

3. The bacteria metabolize the compound to the true mutagen, and humans do not. This seems relatively unlikely, as humans have a wider variety of metabolic and detoxifying systems for foreign compounds than enteric bacteria, though it shouldn't be ruled out completely.

4. The human repair systems are more efficient for a particular type of damage. Not much is known about repair systems in humans as contrasted with bacteria. It is clear that they exist in all systems and are very important, but it seems likely that some small fraction of DNA damage escapes repair in every system.

As repair systems are composed of proteins, there may well be differences between the bacterial and human repair systems in terms of the compounds that can inhibit repair. Caffeine, for example, is a potent inhibitor of repair in bacteria and, though not really mutagenic in the usual sense, is a powerful potentiator of certain mutagens. It would be comforting to know that it is not so in humans (see A4).

5. Chromosomes with histones (eucaryotic cells) are less susceptible to some compounds than chromosomes without histones (bacterial cells). This is hard to evaluate though this may not be very important, as bacterial DNA is probably neutralized with polyamines and eucaryotic DNA may not have histones on all of it or at all times.

## References

Ames, B. N. (1971). In Chemical Mutagens, Principles and Methods for Their Detection, Vol. 1, A. Hollaender, editor. New York, Plenum Publishing Corp.

Ames, B. N. and H. J. Whitfield (1966). Cold Spring Harbor Symposium Quant. Biol., 31, 221.

Craig, A. M. and I. Isenberg (1970). Proc. Nat. Acad. Sci., 67, 1337.

Ficsor, G. and E. Muthiani (1971). Mutation Research, 12, 335.

Hartman, P. E., K. Levine, Z. Hartman, and H. Berger (1971). Science, 172, 1058.

Hollaender, A., editor (1971). Chemical Mutagens, Principles and Methods for Their Detection, Vols. 1 and 2. New York, Plenum Publishing Corp.

Legator, M., and H. V. Malling (1971). In Chemical Mutagens, Principles and Methods for Their Detection, Vol. 2, A. Hollaender, editor. New York, Plenum Publishing Corp.

Malling, H. V. (1971). Genetics, 68, s41 (abstract).

Miller, E. C. and J. A. Miller (1971). In Chemical Mutagens, Principles and Methods for Their Detection, Vol. 1, A. Hollaender, editor. New York, Plenum Publishing Corp.

Miller, J. A. and E. C. Miller (1969). Physico-Chemical Mechanisms of Carcinogenesis. Jerusalem, The Israel Academy of Sciences and Humanities.

Silengo, L., D. Schlessinger, G. Mangiarotti, and D. Apirion (1967). Mutation Research, 4, 701.

Slater, E. E., M. D. Anderson, and H. S. Rosenkranz (1971). Cancer Res., 31, 970.

# THE NEED TO DETECT CHEMICALLY INDUCED MUTATIONS IN EXPERIMENTAL ANIMALS

*Marvin S. Legator*

There can be little doubt that genetic improvement of the human species is based on gradual transformation through the orderly occurrence of mutations that confer a selective advantage to the individuals that carry them. Through the ages, in response to his environment, man has been able to perpetuate the infrequent beneficial mutation and, through the process of natural selection, reject the harmful mutation. This is a continual striving of the species for what has been termed the "adaptive peak."

The mutation rate should never exceed the power of selection to remove harmful genes, for any such increase would lead to genetic degeneration. What factors in our present environment may currently be operating either to increase the production of deleterious mutations or to decrease the removal of harmful mutations by natural selection? The advent of modern medicine, and especially antibiotic therapy, is having a profound influence on natural selection and the prevalence of genetic disease. Individuals with undesirable genetic traits are living longer and reproducing due to improved medical practices. Lederberg recently estimated that 25 percent of our health burden is of genetic origin and that the "genetic legacy" of the species will, in the near future, compete only with traumatic accidents as the major factor in health (Lederberg, 1970).

If the process of natural selection is less active in the removal of undesirable traits than it was 50 or even 20 years ago, what can we theorize about environmental factors that may affect the mutation rate? In the last 30 years there has been a dramatic increase in the number of agents in the environment to which we are all exposed--food additives, pesticides, antibiotics, drugs, and industrial chemicals. Many of these biologically active agents are recent additions to our environment, and hence we are besieged with chemical compounds to which we have not become adapted through the normal process of natural selection. The induction of mutations by various substances has been known since the 1940's, and there can be little question that, among the recently introduced environmental agents, many mutagenic substances are present. It is quite conceivable that, within the last few decades, we have managed to partially neutralize the process of natural selection and simultaneously increase our mutation rate by being exposed to mutagenic agents.

No longer do we have to theorize about chemicals in our environment that could possibly increase the mutation rate in mammals. We have several examples of widely used commercial chemicals that have been experimentally shown to induce mutations in animals. Some pesticides, most notably DDT, were found to induce chromosome abnormalities in cultured mammalian cells and to increase frequency of dominant lethal mutations in rats (Legator, 1970); the widely used agricultural fungicide captan was reported to be mutagenic in several systems (Legator et al., 1969a); prior to the ban of the artificial sweetener cyclamate, because it induces bladder tumor in rats, its metabolite cyclohexylamine was shown to induce chromosome abnormalities in rats (Legator et al., 1969b); and recently, triflupromazine (a widely used phenothiazine tranquilizer) has been shown to produce a dominant lethal effect in rats and mice (see also Drake and Flamm, and Nichols, this volume). Each of these examples are chemicals that we have been exposed to for years--in some instances, for decades. On the basis of structure, these materials cannot be classified as alkylating agents, base analogs, or intercalating agents. They could only be identified as potentially mutagenic agents by suitable laboratory tests in animals. Currently, less than 0.1 percent of all drugs, pesticides, food additives, and industrial chemicals have been investigated for mutagenicity, and virtually no chemical has had the kind of systematic testing that is regarded as adequate.

### The Need to Use Mammalian Systems for Mutagenicity Testing

Since the discovery of chemically induced mutation 30 years ago, scientists have used mutagenic agents as tools to investigate the genetics of many different organisms. Along with this research came the realization that man is also susceptible to chemically induced mutations (Schull, 1962). In the past decade, a series of conferences have been held dealing with the deleterious effects of chemically induced mutations. The high probability that many environmental agents are mutagenic and our inability to effectively monitor the human population in this area have been clearly expressed. The critical problem of eliminating environmental mutagenic agents, however, awaited the development of appropriate methods in mammals.

While there can be little doubt that many fundamental aspects of genetics are best studied in microbes, higher organisms have highly complex metabolic mechanisms that are dissimilar or not even operational in simple systems. A mammalian host can either detoxify or potentiate a specific chemical in a manner different from that of a nonmammalian system, and even the manner of dealing with genetic injury may be dissimilar. For example, most microorganisms readily excise ultraviolet-induced pyrimidine dimers, but certain mammalian cells lack this ability (Fishbein et al., 1970). A significant difference between microorganisms and mammalian cells has also been reported in the fidelity of the translation of the genetic code, with mammalian cells exhibiting far greater exactness than microorganisms (Weinstein, 1968). The need, therefore, to rely on mammalian systems in characterizing mutagenic agents stems not from the differences in reactivity of the mutagen with the genome but rather from the differences in metabolism, the genetic processes dealing with repair, and the fidelity of translation.

In the past few years, relevant methods have been developed to evaluate mutagenic agents in mammals. These methods, along with screening studies in microorganisms, can be used to characterize the majority of mutagenic agents. It should be stressed, however, that the methods are somewhat insensitive, and the possibility exists that some chemicals will slip through our coarse sieves. We can now determine genetically induced death, sterility, semi-sterility, and cytogenetic abnormalities, and we can also evaluate the potential of a chemical to induce point mutations. However, practical methods have yet to be developed that will allow us to determine the detrimental effects in the heterozygous state (recessive mutations), or even dominant or semi-dominant non-lethal mutations. Methods are needed that will measure the results of a large number of concurrent mutations, since ordinarily only small experimental populations are available for study, and our statistical resolving power is extremely limited when we follow only a few selected mutation sites. The evaluation of genetically induced alteration of behavior or intelligence pattern, or histocompatibility studies may be fruitful subjects for the investigation of improved methods in this area. It is fortunate that the known mutagenic agents which produce point mutations at one concentration usually produce effects that we can measure (dominant lethal, sterility, semi-sterility, cytogenetic) at a higher concentration. Present methods will probably detect many mutagenic compounds, but improvement in the detection of point mutations is needed.

The large number of untested environmental agents makes it mandatory that priorities be established for evaluating potential mutagens. Priorities can be established on the following criteria: (1) extent of human exposure, (2) present knowledge of the genetic effects of these agents, and (3) structural relationship to known mutagens.

## Animal (Mammalian) Test Systems

In a program to characterize potential mutagenic substances in our environment, three procedures can be recommended for use: (1) *in vivo* cytogenetics, (2) the dominant lethal test, and (3) the host-mediated assay. These three procedures are practical, precise, efficient, and relatively inexpensive. They are as relevant to man as procedures currently used in the field of toxicology. The combined results of these three *in vivo* procedures should allow the detection of the majority of mutagenic agents. A fourth test, the specific locus test, cannot be considered as a practical routine screen for evaluating chemically induced mutations, but should be reserved for specialized cases where widespread usage or other factors warrant an expensive long-term testing program.

## 1. CYTOGENETICS

With a cytogenetic test system, we can detect morphological evidence of damage to the genetic material. Some of the obvious advantages are: a large number of species, including humans, can be examined by this method; it can be performed on both in vivo and in vitro systems; the genetic material can be observed directly; and the tests can be accomplished relatively rapidly with limited expense.

Heavy reliance should be placed on *in vivo* cytogenetic testing. Somatic as well as germinal cells can be evaluated. Studies can be conducted with either acute, subacute, or chronic doses of the potential mutagen, and both the parent compound and its metabolic products can be tested. Procedures have been worked out for direct bone marrow preparations as well as spermatogonial analysis (Legator et al., 1969b).

The basic prerequisite for cytogenetic investigations include the following points: (1) availability of cells in active division (either natural cell division as in cultured cells, or by use of a mitogenic stimulating agent such as phytohemagglutinin in leukocyte culture), (2) accumulation of sufficient metaphase plates usually through the use of a mitotic arresting agent, (3) hypotonic swelling to further disperse the chromosomes within the cell, (4) preparation of a cellular suspension and its fixation, (5) slide preparation with rupture of the swollen cells to spread the chromosomes in one optical plane, and (6) photomicroscopy to obtain proper plates for analysis.

At the present time, there is no direct evidence that all morphological chromosome abnormalities are derived from mutations, and one cannot equate chromosome breakage with mutations. However, compounds that produce any type of cytogenetic abnormality have usually been shown to induce point mutations, and we can generally consider the induction of chromosome abnormalities to be an excellent indication of genetic damage. Agents which induce chromosomal rearrangements (such as translocations) that do not affect viability of the cell and can be inherited should be considered true mutagens.

## 2. THE HOST-MEDIATED ASSAY

The host-mediated assay can determine the ability of laboratory animals either to activate or detoxify compounds that might be mutagenic (Legator and Malling, 1971). In this assay, the mammal is administered a potential chemical mutagen and injected with an indicator microorganism in which mutation frequencies can be measured. It is important to note that mutagen and organism are administered by different routes. After a sufficient time period, the microorganisms are withdrawn from the animal and the induction of mutants is determined. The comparison between the mutagenic action of the compound on the microorganism directly and in the host-mediated assay indicates whether the host can detoxify the compound or form mutagenic products. The formation of mutagenic metabolic products from dimethylnitrosamine and cycasin (a plant toxin) have been reported using this procedure.

Indicator microorganisms presently being used in this procedure include the histidine auxotroph of Salmonella typhimurium and Neurospora crassa. In addition to flexibility in selection of indicator organism, almost any laboratory animal can be used. Rats, mice, and hamsters have been successfully utilized. Not only can we compare mutagenic activity between microorganisms and mammals but also among different animal species. It should also be possible to demonstrate correlations between mutagenicity and carcinogenicity in the same or different animals.

The host-mediated assay bridges the gap between testing the effects of a potential mutagen in simple microorganisms and in mammals. However, the host-mediated assay does not indicate the effect of specific chemicals on DNA repair mechanisms of the mammal and only indirectly measures mutation in the mammal.

## 3. DOMINANT LETHAL TEST

Chromosomal damage and rearrangements, such as translocations, can result in nonviable zygotes. Evidence for dominant zygote lethality induced in mammals by X-rays and by chemical mutagens has been obtained embryologically and cytogenetically, respectively. Additional evidence for the genetic basis of dominant lethality is derived from the associated induction of sterility and heritable semi-sterility in $F_1$ progeny of males exposed to X-radiation and to chemical mutagens. Translocations have been cytologically demonstrated in such semi-sterile lines in mice and in hamsters (Rohrborn, 1970).

In this test, the mutagen is administered to male rodents, which are then mated, during the cycle of sperm formation, with groups of untreated females. For mice, the duration of spermatogenesis is approximately 42 days, comprising the following stages: spermatogonial mitoses--6days, spermatocytes --14 days, spermatids--9 days, testicular sperm--5.5 days, and epididymal sperm--7.5 days. Thus, matings within three weeks after mutagen administration represent samplings of sperm exposed to the mutagen during postmeiotic stages, and matings from four to eight weeks later represent samplings of sperm exposed during premeiotic and stem cell stages.

Dominant lethal mutations are directly measured by counting early fetal deaths. They can be indirectly measured by reduction in the number of implanted conceptuses in the uterus, compared with control females.

## 4. SPECIFIC LOCUS TEST

The specific locus test is based on detection of newly induced mutation in seven coat-color and morphologic loci in mice. The newly induced mutations can either be chromosome deletions or point mutations. In this test, male mice homozygous for the dominant traits are given the suspect mutagen and mated with female mice homozygous for the recessive traits. In this way, the occurrence of offspring with recessive characteristics is indicative of mutation or loss in the gene (Ehling et al., 1968).

As noted earlier, the number of animals which have to be used to detect a doubling of mutation frequency is so great that the expense of this test makes it impractical to use as a general screening technique.

These three methods, with the exception of the specific locus test, can be carried out in a wide range of experimental animals; any route of administration can be used; and acute, subacute, and chronic studies can be conducted. The tests are comparatively simple, rapid, and economical. The procedures are as relevant to man as other methods currently used in standard toxicological safety evaluations. Experience to date indicates that these methods should be run concurrently in a comprehensive program for evaluating mutagenic agents. No one procedure can be relied on to determine all types of genetic damage caused by various chemicals with different modes of action. The majority of mutagenic agents, however, can be detected by concurrent use of the outlined procedures.

### Impact of Mutagenicity Screening on Carcinogenicity Studies

The correlation between carcinogenic and mutagenic compounds is strong, since the vast majority of known carcinogens have also been found to be mutagenic in one or more biological systems. As we evaluate mutagenic agents in mammals, we can anticipate a greater correlation than is now apparent. While many agents may be found to be mutagenic but not carcinogenic, adequate testing will, I believe, prove that most carcinogens are also mutagenic.

One of the major reasons for conducting two-year toxicity studies in animals has been to determine if the compounds are carcinogenic. In contrast, the mammalian procedures for characterizing mutagenic agents can be concluded in weeks or months. By these procedures, if a compound is found mutagenic, we will have characterized the material as presenting a public health hazard. Then, unless special or unforeseen circumstances arise, we can reach a decision on whether to prohibit use of the compound. Chemicals that are not shown to be mutagenic should, of course, be evaluated for possible carcinogenic manifestations. If a substantial number of compounds which might be carcinogenic are eliminated by mutagenicity testing from commercial use, the savings in time and money to chemical manufacturers would be substantial.

### Insensitivity of Animal Data

In testing for chronic toxicity, we are dependent on animal data to indicate deleterious effects for man. The same limitations apply to mutagenicity testing (which should be considered a sub-area of chronic toxicity. One of the major factors in extrapolating from animal experiments to man is the insensitivity of animal data.

The number of animals that can be conveniently employed to evaluate potentially toxic agents severely limits the extrapolation of deleterious concentrations from animal to man. For example, if animal and man share the same sensitivity to a specific chemical (e.g., if only 0.2 percent of treated animals show an effect), literally thousands --probably at least 5,000-- animals would have to be used in experiments to establish the existence of an effect. The same low incidence of 0.2 percent for a food additive, pesticide, or widely used drug would mean that hundreds of thousands of people would be affected by the agent. To compensate for the need to test a compound in a limited number of animals and to extrapolate the results to widespread use by humans, we must assume a dose response relationship. The dose used in animal experiments is then increased up to the maximum the animals can tolerate. With carcinogenic agents, where a good deal of experience is available, one can cite studies such as the recently completed investigation in which 140 biologically active pesticides were tested in mice at maximum tolerated doses for the lifetimes of the animals, and less than 10 percent were unequivocally carcinogenic (Mrak, 1969). In studies with neonatal mice, where hundreds of compounds have been tested for carcinogenicity, only a small percent of the agents were positive (S. Epstein, personal communication). In the dominant lethal test for mutation, less than four percent of the compounds tested at maximally tolerated doses were positive. Existing information illustrates that most agents tested in mammalian systems do not produce a genetic response in terms of carcinogenicity and mutagenicity. Those that do produce such a response must be regarded as potential hazards, regardless of dosages employed.

### Definitive Animal Experiments – The Only Practical Means of Eliminating Mutagenic Agents

Many agents in our environment have not been tested adequately for carcinogenicity or teratogenicity. This is especially true of chemicals in use for many years before studies in these areas were required. In mutagenicity, which may be of even greater concern from a public health standpoint than either carcinogenicity or teratogenicity, very few chemicals have been evaluated. Nor do firm requirements exist for including mutagenicity tests as part of standard safety testing protocol.

There can be little doubt that we all are exposed to mutagenic agents in our environment. For this reason alone, we urgently need an immediate program to evaluate--in definitive animal studies--the many chemicals in our food, water, and air, and to eliminate wherever possible any agents that show evidence of potential danger to the public health. In addition to testing agents already present in our environment, we must thoroughly screen all food additives, pesticides, industrial agents, and drugs before they are allowed to be introduced into commerce.

The concept that carcinogenic, teratogenic, or mutagenic studies should be conducted directly in any segment of the human population is to condone studies that would be not only impractical but immoral and unethical as well. Such studies in humans could be avoided by carrying out intelligent, definitive, and exhaustive experiments in animals and by utilizing

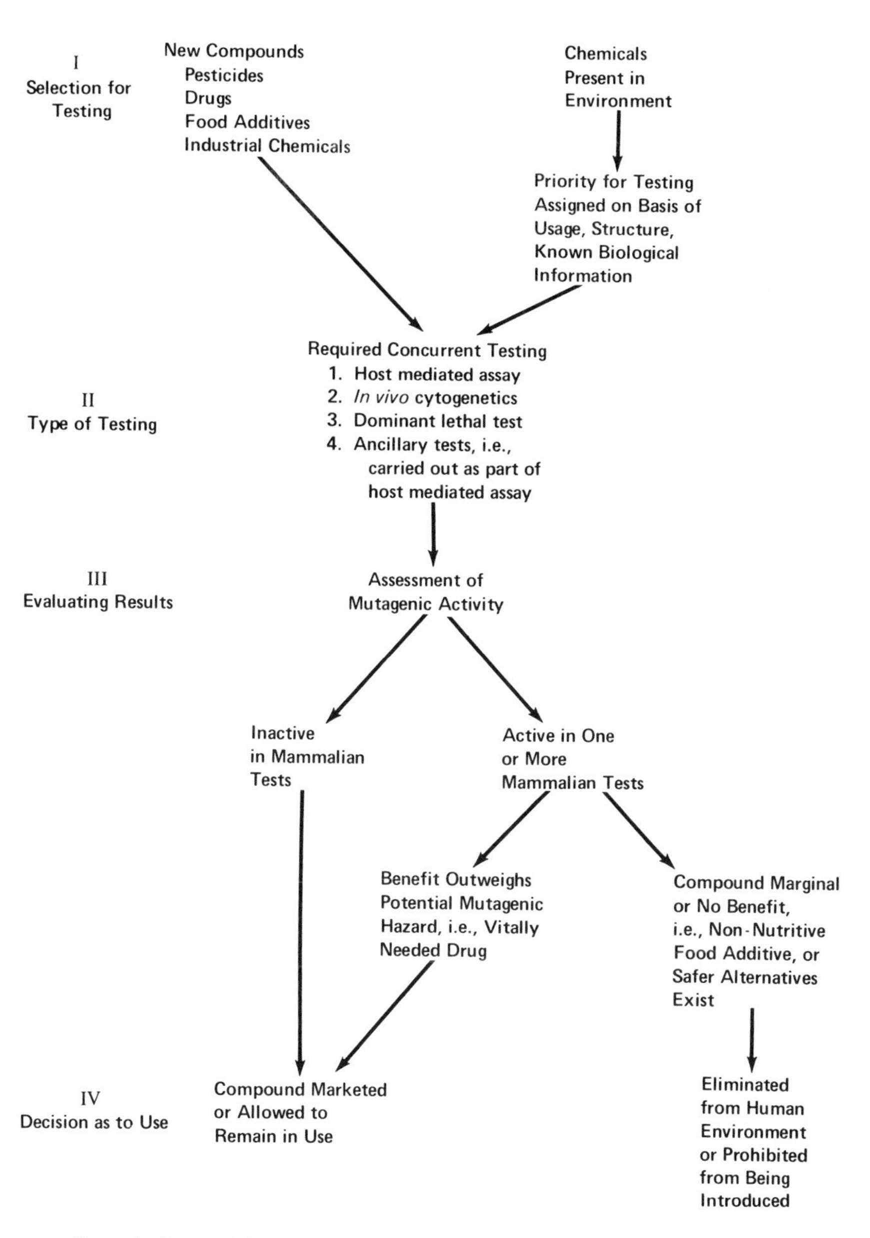


Figure 1. Suggested Scheme for Selecting, Testing and Determining Use of Compound

existing epidemiological data. Figure 1 illustrates a proposed flow sheet for determining if a new material should be introduced into commerce, or if an existing chemical should be allowed to remain on the open market. The three mammalian test systems are used and the final fate of the compound depends upon the benefit/risk ratio. This is essentially the same procedure followed in other areas of toxicology.

Although mutagenic studies should never be conducted directly in the human population, it may be possible to detect harmful agents that are already in our environment by instituting cytogenetic studies and other procedures, e.g., modification of host-mediated assay, in selected individuals such as those occupationally exposed to industrial chemicals.

## References


Ehling, U. H., R. B. Cumming, and H. V. Malling (1968). Mutation Res., 5, 417.

Fishbein, L., W. G. Flamm, and H. L. Falk (1970). Chemical Mutagens, Environmental Effects on Biological Systems. New York, Academic Press.

Lederberg, J. (1970). In Drugs of Abuse: Their Genetic and Other Psychiatric Hazards, S. Epstein, editor. Washington, D. C., U. S. Government Printing Office.

Legator, M. (1970). Pesticides and Chemical Mutagenesis. Symposium, Brookhaven National Laboratory.

Legator, M., and H. V. Malling (1971). In Chemical Mutagens, Principles and Methods for Their Detection, A. Hollaender, editor. New York, Plenum Publishing Corporation.

Legator, M., F. J. Kelly, S. Green, and E. J. Oswald (1969a). Ann. New York Acad. Sci., 160, 344.

Legator, M., K. A. Palmer, S. Green, and K. W. Petersen (1969b). Science, 165, 1139.

Mrak, E. M., Chairman (1969). Report of the Secretary's Commission on Pesticides and Their Relationship to Environmental Health, U. S. Department of Health, Education, and Welfare. Washington, D. C., U. S. Government Printing Office.

Rohrborn, G. (1970). In Chemical Mutagenesis in Mammals and Man, F. Vogel and G. Rohrborn, editors. Berlin, Springer Verlag.

Schull, W. J., editor (1962). Mutations, Second Macy Conference on Genetics. Ann Arbor, University of Michigan Press.

Weinstein, B. I. (1968). In Nucleic Acids, Proteins, and Cancer, Y. Yamamura, T. Aoki, and M. Muramatsu, editors. Japanese Cancer Association, Gann Monograph 4. Tokyo, Maruzen Co., Ctd.

# CHROMOSOME MUTATIONS IN MAN*

*Margery W. Shaw*

## Introduction

Traditionally, human mutation rates have been estimated by attempting complete ascertainment of the sporadic cases of persons with a dominantly inherited phenotype born during a specified time interval in a circumscribed geographical area. Examples include neurofibromatosis, retinoblastoma, Huntington's chorea, multiple polyposis, and aniridia. (For reviews, see Neel, 1962, and Vogel, 1970.) The rates are usually expressed as the number of new mutant genes per locus per generation. The pitfalls of this method are legion: incomplete ascertainment, illegitimacy, phenocopies, reduced penetrance, variable expressivity, and genetic heterogeneity. In addition, the necessary field work and diagnostic confirmation make the method quite tedious and expensive. Although mutation rates of recessive and sex-linked genes have also been estimated, these are considered even less reliable than that of a clear-cut autosomal dominant trait.

With the advent of new laboratory methods for the detection of mutant genes at the molecular level, the enumeration of clinical phenotypes has become obsolete. Other authors in this volume provide details of various screening techniques for estimating the mutation rates of a wide variety of alleles producing biochemical variability. This discussion will be limited to chromosomal mutations and will focus primarily on germinal mutations detected at birth. Particular attention will be given to methods of automation which will eventually make the studies more uniform, more reliable, and less expensive.

---

*Supported in part by USPHS Grant GM 15361.

## Human Chromosomal Polymorphism

It is now well established that the human species pays a large price for chromosomal variability. Approximately one-third of all spontaneous abortions, or five percent of all recognizable conceptions, carry a chromosome abnormality. But nature reduces the number of major chromosomal errors in newborns by ten-fold, or approximately one in every 200 births.

For those mutants which do survive, however, there is not only an economic load imposed upon society but also untold human suffering, both physical and emotional. Fortunately, one-half of these are sterile and thus do not pollute the genetic pool. Nevertheless, even so small a contribution as two per thousand individuals who are new fertile chromosomal mutants is a staggering figure when compared to the estimated mutation rates of specific deleterious genes, which are usually in the neighborhood of $1 \times 10^{-5}$ to $1 \times 10^{-6}$.

It might be argued that chromosomal rearrangements are basically different from "point" mutations and should not be considered in a discussion of the human mutational load. I do not agree, for several reasons: (1) evolution progresses as a result of both mechanisms; (2) there is a wide overlap between mutagens (agents which produce mutations) and clastogens (agents which cause chromosome breakage); and (3) the basic chemical processes are probably quite similar in the two events, i.e., breakage followed by repair. Thus, it is quite likely that induction of chromosomal structural aberrations correlates, in a general way, with damage to intrinsic genes.

## Characterizing Human Chromosomal Mutations

During the past 15 years, great strides have been made in the characterization of the "normal" human chromosome complement, in the identification of specific chromosomes, and in the delineation of chromosome abnormalities. Chromosome aberrations may be classified in various ways, as outlined in Table I. Further discussion of each of these categories is undertaken below, with specific emphasis on the method of detection in each category.

### Monitoring Human Chromosomal Mutations

We are now in a position, technically speaking, to monitor the newborn population and certain high-risk groups (e.g., industrial, chemical, and radiation workers) for increased chromosomal damage due to environmental mutagens. Economically, it is much more feasible to monitor germinal mutations, since this requires observation of only two or three cells per individual to establish the karyotype. There are numerous pitfalls in estimating a detectable increase in chromosome breaks in somatic cells of adults. First, large numbers of cells must be screened on each individual and, ideally, the same individual should be followed over a time span. Thus, it is slow and costly to obtain results. Second, the reliability and repeatability of break rate estimates are less than optimal. Breakage frequencies vary with the observer, with the culture conditions, and with drugs and viral infections in the individual under study. Thus, it is more practical, more feasible technically, more accurate, and more economic to monitor for an increase in germinal chromosomal mutations in the newborn population, especially with automated computer analysis.

Several large scale monitoring programs have been undertaken. Dr. Herbert A. Lubs, of the University of Colorado Medical Center, has kindly

Table I. Classification of Chromosomal Mutations

| | | |
|---|---|---|
| Somatic | vs. | Germinal |
| Autosomal | vs. | Gonosomal |
| Numerical | vs. | Structural |
| <u>In vivo</u> | vs. | <u>In vitro</u> |
| Spontaneous | vs. | Induced |
| Viable | vs. | Lethal |
| Heritable | vs. | Sterile |
| Stable | vs. | Unstable |
| Balanced | vs. | Unbalanced |
| Major rearrangements | vs. | Minor variants |

shared with me some unpublished data on these ongoing projects. Tables II and III give the summaries of five newborn chromosome surveys and five newborn sex chromatin (Barr body) surveys. Although not shown in these tables, it is important to point out that the variation among laboratories is extremely small and not shown to be significant for any of the statistical comparisons which were made. Furthermore, the number of sex chromosome anomalies recovered by the two techniques (karyotypes and sex chromatin determinations) were in extremely close agreement.

Table II. Newborn Chromosome Surveys

| Abnormality | No. Cases | Rate Per 1000 | Approximate Frequency |
|---|---|---|---|
| Sex Chromosome Aneuploidy* | | 2.4 | 1:416 |
| Male | | | |
| XYY | 17 | | |
| XXY | 13 | | |
| XX | 1 | | |
| Female | | | |
| XXX | 8 | | |
| XO | 1 | | |
| Autosome Aneuploidy | | 1.1 | 1:925 |
| Trisomy 13 | 2 | | |
| Trisomy 18 | 2 | | |
| Trisomy 21 | 14 | | |
| Structural Rearrangements | | 1.8 | 1:520 |
| Reciprocal Translocations | 11 | | |
| Centric Fusions (D/D;D/G;G/G) | | | |
| Balanced | 13 | | |
| Unbalanced | 2 | | |
| Other rearrangements | 4 | | |
| TOTAL | 88 | 5.3 | 1:190 |

| No. Tested: | | Laboratories: |
|---|---|---|
| Males | 11,039 | Boston, Massachusetts |
| Females | 5,608 | Edinburgh, Scotland |
| TOTAL | 16,647 | New Haven, Connecticut |
| | | London, Ontario |
| | | Winnepeg, Manitoba |

*Includes mosaics

Table III. Barr Body Surveys (Confirmed by Karyotype)

| | No. Cases | Rate Per 1000 |
|---|---|---|
| Males | | 1.3 |
| XXY | 37 | |
| Mosaic | 11 | |
| Other | 2 | |
| Females | | 1.4 |
| XXX | 30 | |
| XO | 6 | |
| Mosaic | 15 | |
| TOTAL | 101 | 1.4 |

| No. Tested: | | Laboratories: |
|---|---|---|
| Males | 37,771 | Denver, Colorado |
| Females | 35,960 | Geneva, Switzerland |
| TOTAL | 73,731 | Athens, Greece |
| | | Edinburgh, Scotland |
| | | Madison, Wisconsin |

In the ten sections which follow, I shall attempt to provide background cytogenetic knowledge and theory for studying chromosomal mutations according to the comparisons listed in Table I.

## 1. SOMATIC VS. GERMINAL MUTATIONS

A chromosomal change in one of the germ cells of a parent which is passed on to the zygote produces an abnormality that is present in every cell of the offspring. If, however, the zygote receives a normal chromosomal complement and an error occurs after fertilization, then a chromosomal mosaic is produced. Depending on the timing and localization of the error, the mosaicism may or may not be detectable. It must be present in the tissue sampled and it must have occurred at an early stage of development in order to have a reasonable chance of detection. Obviously, a small clone of cells would probably escape notice.

A word of caution should be injected here. If a germinal mutation rate is being estimated, it is not proper to eliminate all chromosomal mosaics on the basis that they are post-fertilization errors. Arguments can be made that most mosaics which are detectable at birth arise in a chromosomally abnormal zygote with subsequent anaphase lag (or chromosomal loss). Several cases of familial mosaicism have been reported and these could be examples of a propensity for somatic chromosome loss in unbalanced cells.

Most chromosomal mosaics have been discovered in numerical aberrations of the sex chromosomes. This is partly due to ascertainment bias since numerical sex chromosome anomalies are, on the whole, more apt to be viable than autosomal anomalies are. But two other reasons for this finding should be considered:

a. A large segment of the Y chromosome is constitutively heterochromatic, and the late-replicating X chromosome is composed of facultative heterochromatin. Heterochromatin in lower animals and plants has been demonstrated to be allocyclic and thus prone to disjunctional errors in meiosis. Cyclic variations of the X chromosome have been demonstrated by special techniques which produce "fuzziness" and heterochromicity in somatic cells (Saksela and Moorhead, 1963). In addition, the appearance of the sex chromatin body in some female interphase cells and a darkly-staining X chromosome in prophase cells (Ohno and Makino, 1961) gives credence to the theory of allocycly. These observations argue for an increased propensity to mitotic errors of the sex chromosomes and would account for the great excess of sex chromosome mosaics.

b. A perusal of the literature of sex chromosome mosaics suggests the possibility that all (or nearly all) of them can be "explained" on the basis of germinal nondisjunction followed by somatic anaphase lag. If nondisjunction occurred only in meiosis, then the sperm or egg would carry from zero to four sex chromosomes. Table IV lists the possible zygotes produced by this hypothesis. All of these have been discovered in man, except, of course, the

Table IV. Numerical Sex Chromosome Abnormalities in Man[1]

| | | EGG | | | | |
|---|---|---|---|---|---|---|
| | | Normal | Abnormal | | | |
| | | X | 0 | XX | XXX | XXXX |
| SPERM: Normal | X | XX | XO | XXX | XXXX[2] | XXXXX[2] |
| | Y | XY | (YO)[3] | XXY | XXXY | XXXXY[2] |
| SPERM: Abnormal | 0 | XO | | | | |
| | XX | XXX | | | | |
| | XY | XXY | | | | |
| | YY | XYY[4] | | | | |
| | XXY | XXXY | | | | |
| | XYY | XXYY[4] | | | | |
| | XXYY | XXXYY[4] | | | | |

[1]All of these can be accounted for by meiotic nondisjunction only.
[2]Zygotes with more than three X chromosomes result from maternal nondisjunction.
[3]The YO chromosomal constitution is lethal.
[4]Zygotes with two Y chromosomes result from paternal nondisjunction.

lethal YO zygote. Furthermore, no other combinations have been discovered (except for three cases of XYYY). In other words, no XXXXXX, no XXYYY, and no XXXXXY cases have been reported. This theory would also account for XX/XY mosaics (when they are not chimeras) and XO/XXX mosaics, on the basis of the loss of a cell line (XXY and XX, respectively). Thus, nondisjunction may be a very rare event in somatic cells, and somatic mosaics may, in fact, be caused by germinal mutations, followed by chromosomal loss but not gain.

The detection of mosaicism is extremely laborious by conventional methods. Errors in counting are very easy to make and it is time-consuming and fatiguing to count and recheck a large number of cells. This type of operation is exactly suited to automation. Once the problem of pattern recognition is solved (and it's close to being solved) there should be no barriers to massive chromosome counts of large numbers of cells per individual, in a search for mosaicism. Many mosaics should be recovered which presently escape detection because of the low percentage of cells in the minor clones and the limitation in the number of cells counted. This problem should be limited only by funds and time in the collection of large numbers of cells for computer counts.

The recognition of germinal mutations by computer is also quite feasible and should theoretically lead to the discovery of more structural rearrangements than would be recognized by the human eye. This will be discussed more fully in Section 3 below.

## 2. AUTOSOMAL VS. GONOSOMAL MUTATIONS

In most cases a chromosomal aberration can be assigned to one of these two general categories. The morphology of the Y chromosome is unique. In preparations of good quality,the Y is distinguishable from the G21-22 autosomes by the absence of satellites (and satellite association) and by its heterochromatic appearance, length difference, and the tendency of the long arms to lie in appositon or parallel rather than flared. Pearson et al (1970) have recently demonstrated fluorescence of the Y chromosome with quinacrine sulfate, while Arrighi and Hsu (1971) have shown differential heterochromatic staining of the distal two-thirds of the long arm of the Y. Autoradiographic identification of the Y chromosome is in doubt (Craig and Shaw, 1971).

Although the X chromosome cannot be differentiated from the C group autosomes by morphology, one of the two X's in female cells is reliably identified by tritiated thymidine autoradiography. Unless it is marked by a morphological variant (Lubs, 1969), it cannot be delineated in male cells.

Among the autosomes, only Nos. 1, 2, 3 and 16 (sometimes 17 and 18) are unique morphologically. In addition, B4-5 and D13-14-15 can be distinguished in some favorably labeled cells. The remainder (groups C6-12, F19-20, and G21-22) have eluded positive identification until recently.

Caspersson et al. (1970) have reported characteristic fluorescence profiles for all of the chromosomes in the human complement. Drets and Shaw (1971) have described specific banding patterns for each of the chromosomes, produced by alkaline hydrolysis followed by incubation in a saline-sodium citrate solution. These newer methods should become quickly adapted to general use and provide far more discrimination in the detection of small chromosomal aberrations which escape notice by conventional cytology.

The computer cannot differentiate autosomal from gonosomal mutations as effectively as the cytologist because it is limited entirely to morphological characteristics. Some programs have been designed, however, to analyze the characteristic banding patterns in _Drosophila_ salivary chromosomes, and this technique should be adaptable to plots of human fluorescent banding produced by the Caspersson technique and heterochromatin banding by the Drets and Shaw technique.

3. NUMERICAL VS. STRUCTURAL MUTATIONS

Chromosomal abnormalities are most frequently classified as numerical or structural. These are thought to be based on independent events, i. e., numerical mutations result from nondisjunction or anaphase lag, while structural mutations result from breakage and reunion.

However, there is increasing evidence that these two classes of mutations are interrelated. There are numerous examples of multiple numerical errors in one individual or among relatives and, more importantly, numerical and structural errors also occur in clusters. Examples include D/D mothers of trisomic mongols, D/G and trisomic mongol sibs, the Philadelphia chromosome with numerical sex chromosome anomalies, and many others. It is also known that clonal evolution and aneuploidy in malignancy occur in cells where structural rearrangements or "marker" chromosomes appear.

Thus, it is becoming increasingly likely that numerical aberrations are based on underlying structural rearrangements. Rather than looking to the spindle fibers or some other extra-chromosomal factor as a cause of nondisjunction, it is quite possible that the behavior of chromosomes during anaphase movement is regulated by the architecture of the entire genome. Interchromosomal effects leading to numerical errors may be based on structural rearrangements.

In attempting to estimate germinal chromosomal mutation rates, it is obvious that numerical errors will almost always be recognized while structural aberrations will more often escape notice--many of them being undetectable with our present methods. Hopefully, automated methods will pick up many which are missed by the human eye. Nevertheless, a large increase in numerical chromosomal mutations would sound a warning signal that structural mutations are also increasing, if the hypothesis of interdependence of the two events is correct.

## 4. IN VIVO VS. IN VITRO MUTATIONS

When LSD was first reported to break human chromosomes _in vitro_, (Cohen et al., 1967) the question of applicability of _in vitro_ tests to _in vivo_ situations, which had been smoldering, broke into a raging controversy. The question has been delineated into two extreme positions: (1) any chemical, virus, or other insult which can be demonstrated to cause chromosome damage _in vitro_ should be suspected of being clastogenic _in vivo_ until proven otherwise; (2) because of the physiology of the organism and the metabolic changes and detoxification which a molecule can undergo _in vivo_, it is not permissible to extrapolate from _in vitro_ tests to _in vivo_ damage.

This question, of course, will not be settled by argument but only by basic research and thorough testing on an applied research level. Several factors should be considered. _In vitro_ testing is a "quick and dirty" method of screening many drugs, food additives, and other chemicals inexpensively. Concentrations can be raised far above toxic levels *in vitro* in order to get a dose-response curve of damage, which would not be permissible in human *in vivo* experiments. *In vitro* tests of human cells may be more enlightening than *in vivo* animal testing because of species differences. Cells which are "exposed" to the external environment, such as the outer layer of cells of the skin, mucous membranes, respiratory tract, and digestive system, may be especially vulnerable to mutagenic agents before *in vivo* detoxification has occurred.

## 5. SPONTANEOUS VS. INDUCED MUTATIONS

Under experimental conditions, the detection of induced chromosome damage is relatively straightforward, since the experimenter knows the agent he is testing, such as radiation, chemicals, or viruses.

Under "natural" conditions, however, a clastogenic agent may be occult or only suspected and a high-risk population is monitored for possible induced mutations. The problems raised here are statistical rather than cytologic. A chromosome breakage rate is determined in the same manner whether a hazard is known or unknown or absent. Only the experimental design will vary (number of individuals, cells per individual, replicate samples, etc.).

Spontaneous chromosome breakage has been estimated in many different laboratories and usually varies from one to five percent of the cells with aberrations, depending on the tissue examined, the culture time, the observer, and the method of scoring. The definition of lesions is far from uniform. The difference between an achromatic gap and a chromatid break is a subjective decision. Restituted breaks will be missed entirely although there may be damage at the molecular level which is biologically meaningful.

Automatic scoring of chromosome breaks is not available. Abortive attempts have been made to write programs for pattern recognition of discontinuity in the chromosome structure, but these are not operational and there is little hope for such automation for several years to come. A rough estimate might be obtained, however, by searching for an increase in the number of chromatin "objects" per cell, counting fragments rather than discarding them as "non-chromosomes".

6. VIABLE VS. LETHAL MUTATIONS

For purposes of this discussion, viable chromosomal mutations will be defined as those in which a live newborn occurs. Many trisomics and polyploids have been described in spontaneous abortuses and stillborns, but these would not be encountered in surveys of live births.

There have been two kinds of newborn surveys: (1) sex chromatin determinations by buccal smear or amniotic cells (for review, see van den Berghe, 1970); (2) actual chromosome determinations done on two to 30 cells per newborn (for review, see Shaw, 1970). Of course, the sex chromatin surveys will miss all of the autosomal abnormalities and some of the gonosomal anomalies (e.g., XYY and sex chromosome mosaics). The chromosome analyses will improve in accuracy with increased cell sample size per newborn achieved by automation.

Tables II and III reveal that the chromosome abnormalities detected in newborn surveys approach 0.5 to 0.6 percent for cytologic analyses and 0.1 to 0.2 percent for sex chromatin determinations. In order to detect a doubling in these baseline values, an extremely large sample size would be required.

To launch such a large-scale newborn monitoring program will require automated techniques. This capability is now at hand and when personnel and money are available, the program could be undertaken (Shaw, 1971).

7. HERITABLE VS. STERILE MUTATIONS

Among common numerical variants, only the XYY, XXX, and occasional trisomy-21 females are fertile, but the first two have only normal offspring. The structural aberrations do not confer complete sterility if they are balanced. Heterozygotes for inversions and reciprocal translocations are able to transmit their rearrangement to offspring although there is "semi-sterility" or increased abortions due to unbalanced gametes. Homozygous rearrangements have not been discovered in man although they have been searched for in racial, religious, and geographic isolates where they might be allowed to flourish. Should they become fixed in a reproductive isolate, the mutation would be reclassified as viable rather than semisterile.

## 8. STABLE VS. UNSTABLE MUTATIONS

A stable chromosome mutation is one which is capable of surviving successive mitotic divisions intact, while an unstable aberration is one which is subject to loss with subsequent death of the daughter cell. Acentric fragments are usually lost due to anaphase lag. Rings and dicentrics undergo the bridge-breakage-fusion phenomenon at anaphase, resulting in loss or gain of chromosome material to one or both daughter cells. Most kinds of rearrangements are stable, however. These include the classical translocations, inversions, duplications, deletions, Robertsonian fusions, and isochromosomes.

Cytologically, unstable rearrangements are easy to recognize since the ring, dicentric, and acentric chromosomes present unique configurations. Although they are not important genetically (since they are non-heritable), they do serve as reliable indicators of the frequency of stable rearrangements which, for the most part, are undetectable. These latter are not detected when the rearrangement is small. A short segment added to a long arm or a slight centromere shift due to an inversion is not detectable by our present cytologic methods, but may become so by automated analysis of banded chromosomes.

## 9. BALANCED VS. UNBALANCED MUTATIONS

A balanced chromosomal mutation occurs when the genes are shuffled around but no loss or gain of genetic material occurs. The individual who is a heterozygote for an inversion or reciprocal translocation suffers no ill effects except, perhaps, sterility or semisterility. In the Robertsonian fusion heterozygote (D/D; D/G; G/G), even though a small centric fragment is lost, there is still no phenotypic effect. Unbalanced mutations include two categories: the primary unbalanced mutations are those such as deletion, duplication, or isochromosome formation, with loss and/or gain of genes, while the secondary type results from a nondisjunctional event in a balanced heterozygote producing an unbalanced offspring. The best example of the latter is the D/G carrier parent who produces a D/G mongol offspring.

## 10. MAJOR REARRANGEMENTS VS. MINOR VARIANTS

So far, we have discussed only the major rearrangements. The minor variants involve primarily the short arms of the acrocentrics and the secondary constriction regions of Nos. 1, 9, and 16, and the Y chromosome. The variation in these regions has been casually observed but difficult to quantitate. It is hoped that heterochromatin staining combined with automated measurement of density will provide more accurate evaluation of these minor variants.

Table V. Minor Variant Polymorphism*

| Variant | Number | Percent |
|---|---|---|
| D short arm excess | 786 | 18.4 |
| D short arm deletion | 2 | |
| G short arm excess | 260 | 6.1 |
| G short arm deletion | 0 | |
| Y long arm excess | 320 | 7.5 |
| Y long arm deletion | 9 | 0.2 |
| No. 1 long arm excess | 11 | 0.3 |
| No. 1 long arm deletion | 0 | |
| No. 16 long arm excess | 103 | 2.4 |
| No. 16 long arm deletion | 22 | 0.5 |
| C inversion (metacentric C) | 12 | 0.3 |
| F inversion (non-metacentric F) | 1 | |
| Y inversion (metacentric Y) | 2 | |
| Total | 1528 | 35.7 |

*Data from New Haven study based on two cells per newborn; 4,283 newborns tested (Herbert A. Lubs, personal communication.)

Table V gives an estimate of the frequency of these minor variants in the newborn survey data of Lubs. Obviously, some of these variants may be categorized as "polymorphic" in the general population. Isolated family studies show them to be heritable, but large-scale genetic studies are yet to be done. We have no idea of the frequency of newly-arising minor variants, i.e., mutations, in this category.

A central question, which has not yet been answered, is whether minor variants are benign or whether they carry some deleterious effects. Starkman and Shaw (1967) showed a higher frequency of D and G polymorphisms in mongols than in controls. Thus, they may tend to increase the risk of nondisjunction. Lubs obtained suggestive evidence that major congenital anomalies and/or low birth weight were positively correlated with enlarged satellites on a G chromosome.

### Detecting an Increase in Human Chromosomal Mutations

We are now ready to monitor the human newborn population for an increase in the germinal chromosomal mutation rate. If we wish to detect a doubling of the chromosomal abnormalities (from 0.5 to 1.0 percent) with 95 percent chance of success and a one percent alpha-type error, we will need a sample of 7,500 newborns. More ambitious, however, would be to begin to think along the lines of total karyotype ascertainment *in utero* by amniocentesis, because detection followed by abortion of abnormal fetuses would be economical over the long run.

Table VI gives the human chromosomal load in terms of number of individuals born per year, rather than rates per 1000. Assuming that the studies in European and American populations are representative, we can estimate that more than one-half million babies with a chromosomal defect are born each year in the world. If we contemplate the control of chromosomal disease and the detection of an increase in chromosomal mutations, this is an area where we are technically ready to begin. The scientific groundwork has been laid; only motivation, money, and personnel are needed to do the job.

Table VI. Human Chromosomal Load

| | U.S. | WORLD* |
|---|---|---|
| POPULATION | 203 million | 3.68 billion |
| NO. ANNUAL BIRTHS | 3.6 million | 124 million |
| BIRTHS PER YEAR: | | |
| Total Chromosome Abnormalities | 19,100 | 656,000 |
| Sex Chromosome Aneuploidy | 8,700 | 298,000 |
| Autosome Aneuploidy | 3,900 | 134,000 |
| Structural Rearrangements | 6,500 | 224,000 |
| Down's Syndrome | 3,000 | 104,000 |
| XYY Syndrome | 3,700 | 127,000 |

*Assuming world rates are equivalent to U.S., Canada, and Scotland rates.

References

Arrighi, F. E. and T. C. Hsu (1971). Cytogenetics, 10, 81.
Caspersson, T., L. Zech, and C. Johansson (1970). Exp. Cell Res., 60, 315.
Cohen, M. M., M. J. Marinello, and N. Back (1967). Science, 155, 1417.
Craig, A. P. and M. W. Shaw (1971). Chromosoma, 32, 364.
Drets, M. E. and M. W. Shaw (1971). Proc. Nat. Acad. Sci., 68, 2073.
Lubs, H. A. (1969). Amer. J. Human Genet., 21, 231.
Neel, J. V. (1962). In Methodology in Human Genetics, W. J. Burdette, editor. San Francisco, Holden-Day, Inc.
Ohno, S. and S. Makino (1961). Lancet, 1, 78.

Pearson, P. L., M. Borrow, and C. G. Vosa (1970). Nature, 226, 78.
Saksela, E. and P. S. Moorhead (1963). Proc. Nat. Acad. Sci., 50, 390.
Shaw, M. W. (1970). Annual Review of Medicine, 21, 409.
Shaw, M. W. (1971). In Perspectives in Cytogenetics, Stanley W. Wright, editor. Springfield, Illinois, Charles C. Thomas Co.
Starkman, M. N. and M. W. Shaw (1967). Amer. J. Human Genet., 19, 162.
van den Berghe, H. (1970). Science, 169, 1318.
Vogel, F. (1970). In Chemical Mutagenesis in Mammals and Man, F. Vogel and G. Rohrborn, editors. Heidelberg, Springer-Verlag.

# THE DETECTION OF INCREASED MUTATION RATES IN HUMAN POPULATIONS*

*James V. Neel*

## Introduction

This presentation will assume that a sufficient case has already been made for studying human mutation rates--either spontaneous, as a basic parameter of human genetics, or induced, as an indicator of the effect on the germ plasm of a wide variety of noxious agents to which human populations are increasingly exposed. With respect to induced mutations and their possible deleterious effects, the early concerns of twenty to forty years ago were, of course, primarily directed toward radiation, with the late Professor H. J. Muller their most eloquent spokesman. More recently, although interest in radiation effects remains very real, the subject of chemical mutagenesis has moved center stage. Detailed reasons for the concern for man on this point will be found in a number of recent publications (1-9) as well as the current volume. In addition, the bulletins of the newly organized Environmental Mutagen Society are quite instructive. A partial list of the chemicals now in our environment which, for one reason or another, could be mutagenic for man includes the fungicide, captan; the plant-growth inhibitor, maleic hydrazide; the

---

*Copyright by the University of Chicago. This paper is an extension of a presentation at the Conference on Genetic Disease Control, held in Washington, D.C., December 3-5, 1970, subsequently published in Perspectives in Biology and Medicine, vol. 14, pp. 522-537, University of Chicago Press (1971). Support of Atomic Energy Commission Grant No. AT(11-1)-1552 is acknowledged.

artifical sweetener, cyclamate; the food preservatives, sodium nitrite and sodium nitrate; certain of the streptomyces-derived antibiotics; various insect chemosterilants, such as triethylene phosphoramide and triethylene melamine; pesticides, such as the mercurials, the carbamates and thiocarbamates, the organophosphates, and the unsaturated "rings" with -OH or -SH groups; the benzepyrine found in smog; and a variety of alkylating agents. There seems no reason to belabor the possibility that human exposures to radiation and chemical mutagens may be increasing mutation rates; the problem to be considered here is how we proceed to evaluate this possibility.

A voluminous literature on the genetic effects of radiation on experimental organisms has already accumulated, and an equally voluminous literature on chemical mutagens is building up rapidly. There is, of course, absolutely no question that experimental models are indispensable in screening potential mutagens and that the results of these screens will, in many instances, be adequate to forestall public exposure to a variety of agents. However, a potential mutagen could pass reasonable screening tests, and yet, alone or in combination with other agents, be mutagenic in man. Furthermore, some agents could have such beneficial effects as pesticides or herbicides that--as in the case of radiation--some genetic risks may be accepted in return for the benefits conferred. Finally, we may inadvertently introduce unrecognized mutagens into the environment. Thus, it is clear that no matter how extensive the screening programs may become, the final test will be what is happening to the germinal tissue of human populations. It will be the thesis of this presentation that the technological advances of the last twenty years permit new and much more refined approaches to monitoring human populations than have been true in the past.

## The Approaches to Monitoring Human Populations

There are three principal approaches to detecting an increased mutation rate in human populations.

## 1. THE USE OF POPULATION CHARACTERISTICS

This approach utilizes as indicators of an increase (or decrease) in mutation rates changes in such parameters of human populations as stillbirth frequency, birth weight, frequency of congenital defect, the sex ratio, death rates during the early years of life, and physical growth and development during infancy and childhood. The argument that an increased mutation rate should be reflected in these indicators is simple and solid. However, the relationship between any given change in the indicator and the underlying change in mutation rate is not clear. Furthermore, we know that all these indicators are influenced by many exogenous factors operating on both mother and child. A suitable control population must be available, and, especially as it concerns the matter of chemical mutagenesis, establishing such a control group becomes increasingly difficult.

This was the approach employed in the principal opportunity available to date to search for evidences of an increased mutation rate in man, involving the children born to the survivors of the atomic bombings at Hiroshima and Nagasaki. It was felt that the population base was not adequate for approach 2 (to be discussed below), and approach 3 was not yet on the scientific horizon. The study, its findings, and the problems in interpreting the results have been presented in several summary publications (10-13). Without undertaking a detailed critique here, let me say simply that in my opinion this approach is superseded by recent scientific developments.

## 2. THE USE OF SENTINEL PHENOTYPES

This approach monitors a defined population for changes in the rate of occurrence of isolated cases (within the family) of certain phenotypes which have a high probability of being due to a dominant mutation. Table I enumerates the majority of the phenotypes which have been utilized in the past to reach estimates of mutation rates in man as well as the estimates so derived. There has been considerable discussion as to how typical these rates are of all human genes. On the one hand, it has been suggested, most strongly by Stevenson and Kerr (14),

Table I. Some Spontaneous Mutation Rates in Man*

| Character | Method of Estimation# | Mutations Per Gene Per Generation | Remarks |
|---|---|---|---|
| DOMINANT GENES: | | | |
| Epiloia........................ | Direct | $(0.4-0.8)X10^{-5}$ | ........................................ |
| Chondrodystrophy............. | Direct | $4.2X10^{-5}$ | Estimates may be spuriously high |
| | | $4.9X10^{-5}$ | because of some evidence for |
| | Direct | $7\ X10^{-5}$ | occurences of phenocopies (15) |
| | Indirect | $4.3X10^{-5}$ | |
| Pegler's nuclear anomaly...... | Direct | $2.7X10^{-5}$ | ........................................ |
| Aniridia...................... | Direct | $0.5X10^{-5}$ | ........................................ |
| Retinoblastoma................ | Direct | $1.4X10^{-5}$ | Estimate based on assumption that all |
| | | $2.3X10^{-5}$ | sporadic cases are due to mutation |
| | | $4.3X10^{-6}$ | Estimate based on proposition that approximately 75% of all sporadic cases are phenocopies |
| Waardenburg's syndrome........ | Direct | $3.7X10^{-6}$ | ........................................ |
| Neurofibromatosis............ | Direct | $(1.3-2.5)X10^{-4}$ | ........................................ |
| | Indirect | $(0.8-1.0)X10^{-4}$ | ........................................ |
| Facio-scapulo-humeral progressive muscular dystrophy.................. | Direct | $4.7X10^{-6}$ | ........................................ |
| | Indirect | $4.7X10^{-6}$ | ........................................ |
| Multiple polyposis of the colon...................... | Indirect | $(1-3)X10^{-5}$ | ........................................ |

| | | | |
|---|---|---|---|
| SEX-LINKED RECESSIVE GENES: | | | |
| Hemophilia.................... | Indirect | $3.2\times10^{-5}$ | Estimate may include 3 distinct types of hemophilia: classical sex-linked hemophilia resulting from deficiency of antihemophilic globulin, a sex-linked clotting defect from lack of "plasma thromboplastin component," and an autosomally inherited clotting defect from lack of "plasma thromboplastin antecedent" |
| Childhood progressive muscular dystrophy.......... | Direct | $3.2\times10^{-5}$ | ..................................... |
| | | $1\times10^{-4}$ | Not a true direct estimate but an approximation which overestimates the mutation rate |
| | Indirect | $3.8\times10^{-5}$ | ..................................... |
| | | $(4.5\text{-}6.5)\times10^{-5}$ | ..................................... |
| | | $1\times10^{-4}$ | ..................................... |

SOURCE.--Neel and Reed. In:P. Altman and D. Dittmer, eds. Growth, p.103. Washington, D.C.:Federated Societies, 1962.

*Knowledge of mutation rates in man, as in other organisms, is still considered provisional, and the traits for which estimates are available are a highly select fraction of all genes. Estimates for chondrodystrophy, aniridia, and neurofibromatosis may be more reliable than those for epiloia and Waardenburg's syndrome.

#Estimates are considered to be direct when based on observed mutations; indirect when not so based. All direct estimates make use of determinations of the relative fitness and frequency at birth of the trait and assume that the population is in equilibrium.

that since one does not select the very rare inherited phenotypes for studies of mutation rate, the existing estimates involve in general the more mutable genetic loci. From a careful survey of (recessive) sex-linked traits, they suggest that the average rates are more nearly $1 \times 10^{-6}$. The use of recessively inherited traits in mutation rate estimates is plagued by the question of heterozygote effects. On the more theoretical level, it can be argued that all loci have very similar mutation rates, but that for some any change in the gene product has a phenotypic counterpart, whereas other loci are much less "sensitive" (15). If this is so, then, although there is certainly an element of bias in the selection of traits for such studies, the existing estimates are not so unrepresentative as they appear on first study. I find it thought provoking how similar the average of the human estimates to date, about $2 \times 10^{-5}$ (or $1 \times 10^{-5}$, if neurofibromatosis is omitted), is to the average for eight loci in the mouse ($0.75 \times 10^{-5}$; (16)). The mouse loci, although selected for ease of mutation detection, were not selected for spontaneous mutation rates. If one views this average human rate as a gross overestimate, then one is forced to the position that human rates of mutation are lower than those of mice! However, for monitoring purposes, it is relatively immaterial whether these rates are typical, providing the response to mutagens of the loci involved is characteristic of the genome as a whole.

But there are other problems, theoretical and practical, in the use of sentinel phenotypes. On the theoretical side, isolated cases of some of these phenotypes can possibly represent phenocopies, the result of somatic mutation, or can be due to recessive inheritance. Mutation at any one of several different loci may result in clinically indistinguishable phenotypes. In laboratory jargon, these are not "clean" systems. On the practical side, the chief problems are selecting traits for which at an early age there exist minimal diagnostic problems, and then establishing a suitable registration system in a sufficiently large study population. Let us consider the size of the population which must be under intensive surveillance for the occurrence of these defects in order to detect -- let us say -- a 50 percent increase in mutation rates in a single year, which seems the maximal increase that should be permitted to go undetected. Let us assume that those phenotypes listed in Table I which would be suitable for such a study (and possibly several others not yet recognized) collectively amount to 50 per 100,000 births. We do not have a theoretical rate, only an estimate. Our question is, then, What is the size of the population necessary to demonstrate at the 0.05 or 0.01 level of significance an estimated 50 percent increase in mutation rates? In other words, how many births must we monitor to be confident, at the indicated statistical level, that a change from an estimated 50 per 100,000 births to 75 per 100,000 births has occurred? The computation must take into account Type II errors, that is, the probability of rejecting a real change in rate of the magnitude indicated (17). The numbers shown in Table II are based on a one-tailed probability test. If we accept a $\beta$ value of 0.20, this requires two samples of 127,000 births at the 0.05 level, and two samples of 206,000 at the 0.01 level. At a crude birth rate of 20 per 1000 population, this entails a study population of either 6,337,500 or 10,283,000 if one wishes to put the study on a year-to-year basis.*

## 3. THE USE OF BIOCHEMICAL AND/OR CHROMOSOMAL MUTATIONS

It is traditional to distinguish between two types of mutations, namely "point", in the strict sense involving alterations in the genetic code, and "chromosomal", usually thought of as involving chromosomal additions, subtractions, or structural alterations detectable with the light microscope. Recent and pending developments in human cytogenetics and human biochemical genetics raise the possibility of examining mutation in man at both these levels on a scale and with a precision unthinkable only a few years ago. The subject of the detection of chromosomal mutations falls outside my charge, and I will say no more on that score (see Shaw, this

---

*Since this article was submitted, F. Vogel (in Chemical Mutagenesis in Mammals and Man, F. Vogel and G. Rohrborn, eds., p. 445, Berlin, Springer Verlag) has published very similar calculations.

Table II. The Sample Sizes Necessary to Detect an Increase in Certain Sentinel Phenotypes#

| Value of α | Value of β | | |
|---|---|---|---|
| | .10 | .20 | .51 |
| .05 | 176,000 | 127,000 | 54,000 |
| t-test | 2.96*** | 2.52** | 1.65* |
| .01 | 267,000 | 206,000 | 108,000 |
| t-test | 3.65**** | 3.21*** | 2.33** |

#From an observed 50 per 100,000 births to 75 per 100,000 births, at the indicated probability levels for Type I (α) and Type II (β) errors. The *t*-values are based on one-tailed probability distributions.
*$0.05 > P > 0.01$
**$0.01 \geq P > 0.005$
***$0.005 \geq P > 0.0005$
****$0.0005 \geq P$

volume). With respect to "point" mutations: formerly they were studied at the level of gross organismic phenotypes, the genes for which could by suitable breeding tests be localized to specific regions in chromosomes. In man, the ability to study point mutation in the traditional sense remains seriously restricted by the lack of suitable chromosomal markers, the small size of the kindreds generally available for study, and the length of the human life cycle. Additional difficulties have been mentioned in the discussion of "sentinel phenotypes", which, until proven otherwise, are assumed to result from point mutation. However, recent developments in our ability to isolate and characterize a wide variety of human proteins, coupled with the knowledge of the genetic code and its relationship to amino acid sequences, would seem to permit us to circumvent many of those previous difficulties and approach the subject of mutation in man at a new and far more precise level. In short, it is now possible to examine a representative series of proteins in placental cord blood or blood obtained during the first year of life for variation in either serum or erythrocyte proteins; and when a variant is encountered, to determine by suitable family studies whether it has been inherited from one or the other parent, or whether both parents are normal with the variant presumably resulting from mutation. The choice of material is dictated solely by technical considerations, so that there should be none of the possible bias discussed above with reference to mutability. Likewise, the problem of phenocopies and recessive inheritance contributing to the phenotype and so distorting mutation rate estimates seems minimal with this approach.

It seems clear that this last approach, still untried, in theory offers a more precise evaluation of the occurrence of a change in mutation rates than do the other two. There are, however, a number of questions which must be resolved before any large-scale venture into this field seems indicated. Since the pros and cons of the other two approaches have been discussed in some detail during the past twenty years, we propose to devote the remainder of this presentation to an examination of the feasibility of approach 3.

## Some Problems in Monitoring for Mutations at the Chemical Level

It is convenient to divide the questions raised by the biochemical approach into "primarily technical" and "primarily theoretical".

### 1. PRIMARILY TECHNICAL QUESTIONS

#### a. Choice of Indicators

Table III lists, with no attempt at justification, a series of proteins in which inherited electrophoretic variants, proven or presumably due to single amino acid substitutions, have been detected. An attempt has been made at a division into "good" and "less useful" proteins, based on current data on the ease of detection of variants. Typing for some of these systems is

Table III. Blood Proteins Suitable for Screening

| Serum | Erythrocyte |
|---|---|
| | GOOD |
| $\alpha_2$-macroglobulin | Acid phosphatase |
| Albumin | Adenosine deaminase |
| Pre-albumin | Adenylate kinase |
| Pseudocholinesterase | Carbonic anhydrase |
| Transferrins | Catalase |
| | Diaphorase |
| | "Esterases" |
| | Galactose-1-phosphate uridyl transferase |
| | Glucose-6-phosphate dehydrogenase |
| | Glyceraldehyde-3-phosphate dehydrogenase |
| | Hemoglobin |
| | Hexokinase |
| | Isocitrate dehydrogenase |
| | Lactic dehydrogenase |
| | Leucine aminopeptidase |
| | Malic dehydrogenase |
| | Peptidase A |
| | Peptidase B |
| | 6-phosphoglucose dehydrogenase |
| | Phosphoglucomutase |
| | Phosphohexose isomerase |
| | LESS USEFUL |
| Ceruloplasmin | Aldolase |
| Haptoglobin | Glutamate dehydrogenase |
| | Glutathione reductase |
| | Peptidase C |
| | Pyruvic kinase |

difficult in the very young. The list will surely need revision, and additions may be anticipated in the near future. However, the table does document the statement that even now there are at least twenty good indicators, with the possibility that several on the "less useful" list might, with improved techniques, move into the "good" category. In passing, we recognize the possible occurrence in all these systems of electrophoretically silent mutations, that is, amino acid substitutions which do not result in a change in molecular charge. Their frequency can only be calculated accurately when the precise amino acid composition of the protein is known; for hemoglobin, more than half of the theoretically possible substitutions are of this type. It would be an assumption that the proportion of such mutants is the same for induced and spontaneous mutation; within limits, this assumption is subject to investigation.

b. Multisystem vs. Single System Methods

Many laboratories searching for genetic variants in serum proteins employ "multisystem" methods. For instance, in our laboratory we currently search for haptoglobin, ceruloplasmin, transferrin, and Gc variants by introducing several drops of serum into a starch gel, followed by 600-volt electrophoresis at pH 8.1-8.3 for sixteen hours, after which the gel is split longitudinally into three strips, one stained with benzidine-sodium nitroprusside for haptoglobin, the second with dianisidine for ceruloplasmin, and the third stained with amido black, from which we read transferrins and Gc. Likewise, acid phosphatase, phosphoglucomutase, lactic dehydrogenase, and malic dehydrogenase are all run on the same gel at pH 7.0. This procedure to some extent represents a compromise with the best method to demonstrate variants of any single protein, the final decision as to pH and duration of electrophoresis being based on how well the variants which occur in polymorphic frequencies are demonstrated. But if we are engaged primarily in a search for new mutants, perhaps each protein should be examined separately in a system optimum for that protein. Furthermore, since there are examples of an apparently optimum system failing to demonstrate a variant easily detected at another pH, perhaps we need to screen each protein for variants at several different pH's. In short, assuming that the results of these extensive studies will provide the most definitive data thus far concerning the frequency of protein variants, it is clear there is a need for considerable thought concerning the optimum methodology. A consideration which will surely enter in at some point will be the need to conserve serum.

c. Automation

Spectacular advances have recently been made in the automation of a variety of biochemical determinations involving blood serum, especially by Dr. Norman Anderson and his group at Oak Ridge (18,19). These determinations for the most part involve quantitative traits, and it is by no means clear that the qualitative differences with which the search for mutants in proteins is concerned will yield to the same approach. Nevertheless, electrophoresis has so many of the characteristics of a cottage industry that if truly large-scale work is contemplated, clearly time spent on automation might be time well spent.

d. The Complication of the Genetic Polymorphisms

In many of the systems which it will be convenient to study, some variants occur in polymorphic proportions (greater than 2 percent of the population), whereas others are essentially "private", restricted (thus far) to one or several families. For the most extensively studied protein to date, hemoglobin, there are three major polymorphisms (S, C, and E) plus private variants whose collective frequency has been found to be approximately 1 in 2000 in Europeans (20) and 1 in 5000 in Japanese (21). It will not be feasible to do family studies on each child exhibiting a polymorphic variant, such as hemoglobin S, on the remote possibility that this might represent a new mutant to this gene. Unfortunately, variants are known whose electrophoretic mobility is similar to that of a well-established polymorphism and which would in the usual screening procedure be grouped with the polymorphism. For instance, there are at least thirty different hemoglobin variants with the mobility of hemoglobin S, all of which would, without supplementary procedures, be assumed to be hemoglobin S in a population with a high frequency of the sickle cell trait. Although in the case of hemoglobin S one could introduce a simple supplementary procedure such as a solubility study, such simple tests are not always available.

e. Distinguishing between Mutation and Discrepancy between Legal and Biological Parentage

In the United States, approximately 1.5-3.5 percent of Caucasian children born in wedlock present, on the basis of genetic markers, discrepancies between legal and biological parentage (22,23). How will this fact disturb studies of human mutation rates? The problem may be approached in several different ways; perhaps the simplest is as follows: the ability to detect these discrepancies depends, of course, on the number of genetic systems employed and the gene frequencies of these systems. Race and Sanger (24) calculated that, employing only seven systems at the gene frequencies of English populations, the probability of detection of non-paternity was 60 percent. I have not attempted an exact calculation for the approximately twenty systems now available (including the HLA system). Let us assume we are at the 90 percent plus level. Then, where apparent mutation is due to nonpaternity, for every nine where there is demonstrable non-paternity, with the presumption the gene was introduced by the unknown parent, there will be one undetected case for whom allowance must be made in any calculation of mutation rates. In effect, by a suitable adjustment of the denominator, the estimate is based on children whose biological parents are known at a high level of accuracy.

If the mutational event is sufficiently rare, undetected nonpaternity could conceivably dominate a study of mutation rates. Let

$I$ = frequency of illegitimacy, assumed $5 \times 10^{-2}$

$X$ = frequency in the population of genes responsible for rare private variants, assumed $2 \times 10^{-4}$

M = mutation rate (detectable) to the rare gene, assumed, to be conservative, $0.5 \times 10^{-5}$

Then

$$\text{Apparent mutants} = M(1-IX) + (1-M)IX + MIX$$

$$\simeq M(1-IX) + (1-M)IX$$

$$\simeq 0.5 \times 10^{-5} + 1 \times 10^{-5}$$

This calculation suggests that any group of apparent mutations will be comprised of mutants and "paternity exceptions" in equal numbers. Since, however, we can exclude 90 percent of the latter, we will still have a valid study of mutation. If, however, mutation rates were actually an order of magnitude less, and discrepancies of the type under discussion increased to 10 percent, then our putative study of mutation would be dominated by events resulting from nonpaternity. The very preliminary data from *Drosophila* suggest that the spontaneous rate of mutation of loci coding for protein structure may be less than the commonly accepted rates based on gross phenotypes of the order of $10^{-7}$ (25), so that the paternity problem might be non-trivial. (This section has profited from discussions with Dr. James Crow and Dr. Robert Krooth.)

### f. Sample Size

By the standards of the past, this approach requires a major effort, both for the original screenings and the follow-up studies on persons found to possess variants, to separate mutants from nonmutants. Let us assume that the rate at which spontaneous mutation results in detectable changes in a protein is $0.5 \times 10^{-5}$ per locus per generation. Let us assume we can monitor for 20 different proteins (some associated, like the $\alpha$ and $\beta$ chains of hemoglobin) and that we wish to be able, at the 0.05 and 0.01 probability limits, to detect a 50 percent increase in the frequency of mutation. We wish to detect an increase from an estimated 20 per 100,000 births to 30 per 100,000 births. If we accept a Type II ($\beta$) error of 0.20 this requires at the 0.05 level 365,000 person-determinations in each sample or, at the 0.01 level, 592,000 person-determinations. With automation, 365,000 person-determinations represents the annual output of perhaps 40 technicians and 5 supervisory scientists. A Type II error of 0.20 is fairly restrictive, but I have made the calculation at this level on the assumption that one wants to be reasonably certain of not missing a real effect. If the contrasts are to be made annually, employing cord blood, the base population, at a crude birth rate of 20 per 1000, must be approximately 18,256,000 or 29,622,000 persons. The necessary numbers for different values of β are given in Table IV. It is apparent that not only a large laboratory but a formidable "collecting net" must be organized. With a study of this magnitude, the decision to proceed must be considered with great care. However, geneticists, having emphasized the risks to society of increased mutation rates, can scarcely draw back now from the effort to assign this undertaking a proper priority in society's needs.

Table IV. The Sample Sizes Necessary to Detect an Increase in Certain Biochemical Mutants#

| Value of α | Value of β | | |
|---|---|---|---|
| | .10 | .20 | .56 |
| .05 | 506,000 | 365,000 | 135,000 |
| t-test | 3.18*** | 2.70*** | 1.65* |
| .01 | 768,000 | 592,000 | 270,000 |
| t-test | 3.92*** | 3.44**** | 2.33** |

#From an observed frequency of 20 per 100,000 births to 30 per 100,000 births, at the indicated probability levels for Type I (α) and Type II (β) errors. The *t*-values are based on one-tailed probability distributions.

*0.05 > *P* > 0.01
**0.01 > *P* > 0.005
***0.005 > *P* > 0.0005
****0.0005 > *P*

### g. The Proportion of the Mutation Spectrum Detected

Although the function of enzymes appears to be rather robust in the face of single amino acid substitutions, it is not clear that it would be equally robust to small additions or deletions in the polypeptide chain. Thus, it would appear desirable wherever possible to employ protein stains rather than detection systems relying on enzyme activity. Whether the important class of mutations resulting in absence of enzyme protein can be detected in the heterozygote is debatable. In short, although this approach will greatly extend our knowledge of the frequency of certain types of mutations, it too has limitations.

## 2. PRIMARILY THEORETICAL ISSUES

### a. What is the Impact of an Increased Mutation Rate on the Fitness of Subsequent Generations?

Our concern over the prospects of an increase in mutation rates stems, on the one hand, from the intuitive belief that the mutation rate of a species should, through selection, be close to an optimum for that species, and, on the other hand, from the experimental demonstration, especially in *Drosophila*, that most mutations have deleterious effects. While the consensus would still favor these generalizations, a number of recent developments have weakened their impact. With respect to optimum mutation rates, it can be argued that in this period of rapid transition for our species, in theory a modest increase in mutation rates could facilitate the necessary genetic adjustments. The vexing moral issue immediately raised by this argument is the temporary "price" to be paid for the adjustment, in the form of an increase in individuals with certain genetic defects.

As regards the concept of the generally deleterious effects of mutation, the large amount of previously concealed variability recently revealed by biochemical techniques challenges many of the formulations of population genetics. This challenge is blunted if it is assumed that those "new" alleles resulting from mutation which are retained in populations are often neutral with respect to natural selection (26-28). Furthermore, mice, rats, and swine receiving relatively large doses of radiation over a series of generations do not appear to show net dominantly inherited deleterious effects, suggesting, *inter alia*, an unexpectedly "efficient" elimination of new mutants; or that current estimates of induced mutation rates based on specific locus studies in mice are too high; or that estimates of the average heterozygous deleterious effects of induced mutations in *Drosophila* are not applicable to mammals (29-33). Finally, under certain circumstances, irradiation over a series of generations actually seems to improve the fitness of *Drosophila* populations (rev. in 34). This demonstration does not contravene the principle that induced mutations are in general deleterious, but rather emphasizes the importance of understanding the selective process in long-range predictions. The inference from Wallace's work (34) is that it is the effects of a mutant when heterozygous, rather than when homozygous, that determine its retention in a population. As inbreeding relaxes and human populations amalgamate, the same might be increasingly true for man.

All these developments have made us more cautious in predicting the impact on populations of an increased mutation rate. On the other hand, should the appropriate studies yield no evidence for a greater mutation rate in human populations exposed to higher concentrations of potential mutagens, then the concerns of a large sector of the biological community will be greatly relieved. Incidentally, an advantage of the sentinel phenotype approach is a much better feeling for the impact on human populations of the mutational event one is measuring.

b. *How are "Negative" Data to be Interpreted?*

In discussions of this topic, concern is often expressed that a study which failed to yield statistically significant findings would be interpreted as "negative" and so result in a false reassurance. This danger can be met by a clear recognition from the outset that there is in this type of investigation no such thing as a negative outcome. Rather, the data should be used to set upper limits, with specified probabilities, on the magnitude of the effect which might go undetected. In other words, in a sample of appropriate size, although we might not be able to demonstrate a statistically significant increase, we could exclude, say, the possibility that the mutation rate had increased by 100 or more percent (12).

### c. If an Increased Mutation Rate is Detected, How to Identify the Responsible Agent(s)?

Earlier I mentioned a few of the potential mutagens to which human populations are exposed. Given an increase in mutation rates, the task of identifying which of these potential mutagens are responsible is, of course, formidable. We may never be able to solve the problem by direct observation on human populations but rather, on the basis of experiments with laboratory mammals, attempt to reduce exposure to the most potent mutagens. Whether the real offenders have been identified will quickly become apparent with subsequent monitoring. The necessary decisions will probably never be based on genetic considerations alone but on knowledge of the somatic effects as well, both the genetic and somatic factors balanced against the benefits to society from continued use (hopefully, temporarily) of the agent.

It was considerations of this latter type which led me to entitle a recent presentation on this subject "The Evaluation of the Effects of Chemical Mutagens on Man: Long Road Ahead" (9). Like others of the biological issues upon us, this one will not yield easily. On the other hand, given that our species successfully navigates the present difficult transition period and returns to a "quasi-equilibrium" state such as characterized our remote ancestors (35), then surely concern for the protection and continuing evolution of man's genetic endowment has to be a major factor in how we design that quasi-equilibrium.

## Concluding Remarks

Although this presentation perforce has been directed toward the detection of mutation, it seems appropriate to mention briefly two inevitable concomitants of any large-scale program directed toward the detection of mutation at the chemical level. First, within limits, such a program could also detect certain types of disease. It thus would furnish an experience in early disease detection. Second, the data which would result concerning variation at this biochemical level are of enormous interest to population geneticists.

In closing, then, let me once again emphasize that it now appears feasible to study the important phenomena of spontaneous and induced mutation in man at a new and far more precise level than was previously possible. A major effort is involved. Where to rank this in our hierarchy of genetic priorities will surely be the topic of intensive discussion in the years immediately ahead. In the meantime, our group is exploring the feasibility of initiating, in concert with appropriate Japanese investigators, such a study on the children born to the survivors of Hiroshima and Nagasaki and on a suitable group of controls. This proposal is based on the thesis that not only is this the most significant single group in the world for this type of study and the group to which it would mean the most, but also that the experience so gained would be of great usefulness to all industrialized nations concerned with evaluating the effects of environmental contamination upon their citizenry.

Recent developments raise the possibility of monitoring human blood cells for the occurrence of mutation in somatic cells (36; see also Sutton, this volume). If a defined relationship between somatic and germinal mutation could be established, then it would, of course, be preferable to use the cheaper and quicker techniques of monitoring for somatic cell mutations. Accordingly, it would seem almost mandatory to attempt to detect evidences of an increased rate of somatic cell mutations in the parents of the children being screened for evidences of increased germ line mutations, in a study such as may be undertaken in Hiroshima and Nagasaki. Since studies are already in progress on the cytogenetic implications for children of the exposure of their parents to the atomic bombs (37), the opportunity exists to investigate a three-way correlation. The possible loss of information inherent in beginning such studies so long after the bombings is, perhaps, offset by our concern to know what kinds of defects persist in the $F_1$ and may be transmitted to subsequent generations.

**References**

1. A. Barthelmess. Arzneimittelforschung, 6:157, 1956.
2. A. Goldstein. In: Mutations, W. J. Schull, editor, p. 167. Ann Arbor: University of Michigan Press, 1962.
3. F. Vogel, G. Rohrborn, E. Schleiermacher and T. M. Schroeder. Deutsch. Med. Wschr., 49:2249, 1967.
4. J. F. Crow. Scientist and Citizen, June-July, p. 113, 1968.
5. S. S. Epstein and H. Shafner. Nature, 219:385, 1968.
6. J. V. Neel. In: Proceedings on Senate Res. 78, March 4, 5, 6, April 24, and May 7, 1969. Subcommittee on Intergovernmental Relations of the Committee on Government Operations, U. S. Senate, 1969.
7. J. V. Neel and A. D. Bloom. Med. Clin. North America, 53:1243, 1969.
8. Report of the Secretary's Commission on Pesticides and Their Relationship to Environmental Health. U. S. Dept. of Health, Education, and Welfare. Washington: Government Printing Office, 1969.
9. J. V. Neel. Proc. Nat. Acad. Sci., U.S., 67:908, 1970.
10. J. V. Neel and W. J. Schull. Nat. Acad. Sci.-Nat. Res. Counc., Pub. No. 461., p. 241, 1956.
11. J. V. Neel. Changing Perspectives on the Genetic Effects of Radiation. Springfield, Ill.: Thomas, 1963.
12. H. Kato, W. J. Schull and J. V. Neel. Amer. J. Hum. Genet., 18:339, 1966.
13. W. J. Schull, J. V. Neel and A. Hashizume. Amer. J. Hum. Genet., 18:328, 1966.
14. A. C. Stevenson and C. B. Kerr. Mut. Res., 4:339, 1967.
15. J. V. Neel. In: Effects of Radiation on Human Heredity. Geneva: World Health Organization, 1957.

16. W. L. Russell. In: Repair from Genetic Radiation, F. H. Sobels, editor. London: Pergamon Press, 1963.

17. E. Paulson and W. A. Wallis. In: Selected Techniques of Statistical Analysis, C. Eisenhart, M. W. Hastay and W. A. Wallis, editors. New York: McGraw-Hill Book Co., 1947.
18. N. G. Anderson. Science, 166:317, 1969.
19. ______________. Amer. J. Clin. Path., 53:778, 1970.
20. H. Lehmann and R. G. Huntsman. Man's Hemoglobins. Amsterdam: North Holland Press, 1966.
21. T. Yanase, M. Hanada, M. Seita, I. Ohya, Y. Ohta, T. Imamura, T. Fujimura, K. Kawasaki, and K. Yamaoka. Jap. J. Hum. Genet., 13:40, 1968.
22. L. E. Schacht and H. Gershowitz. Proc. Second Intern. Congr. Hum. Genet., p. 894. Rome, Instituto Mendel, 1961.
23. C. F. Sing, D. C. Shreffler, J. V. Neel and J. A. Napier. Amer. J. Hum. Genet., 23:164, 1971.
24. R. R. Race and R. Sanger. Blood Groups in Man, 4th Ed. Oxford: Blackwell, 1962.
25. K. Kojima. Personal communication.
26. M. Kimura. Nature, 217:624, 1968.
27. ________________. Proc. Nat. Acad. Sci., U.S., 63:1181, 1969.
28. J. L. King and T. H. Jukes. Science, 164:788, 1969.
29. E. L. Green. Radiat. Res., 35(2):263, 1968.
30. J. L. King. Genetics, 58:625, 1968.
31. B. A. Taylor and A. B. Chapman. Genetics, 63:455, 1969.
32. ________________. Ibid, 63:441, 1969.
33. P. D. Mullaney and D. F. Cox. Mutation Res., 9:337, 1970.
34. B. Wallace. Topics in Population Genetics. New York: W. W. Norton & Co., 1968.
35. J. V. Neel. Science, 170:815, 1970.
36. H. E. Sutton. In: Monitoring, Birth Defects, and Environment: The Problem of Surveillance, E. Hook, editor. In press.
37. A. A. Awa, A. D. Bloom, M. C. Yashida, S. Neriishi, and P. G. Archer. Nature, 218:367, 1968.

# MONITORING SOMATIC MUTATIONS IN HUMAN POPULATIONS*

*H. Eldon Sutton*

The need to monitor human populations directly for mutational risk is based on the fact that observations on nonhuman systems, however helpful they may be in predicting human effects, cannot duplicate the unique metabolism of man or the variety of environmental components to which he is exposed. Because of his physiology and metabolism, man may be more or less sensitive to mutation by a specific agent as compared to other model organisms. Further, the potential interactions of environmental agents are far too numerous to test one by one, even if it were possible to identify them. Man is both genetically and environmentally heterogeneous. The ultimate test of whether human beings are subject to excessive mutation must be by direct observation.

Screening for mutations may be done with two rationales. First, the screening may be set up to monitor populations at risk for some identifiable agent or, second, monitoring may be done for a large population over time to detect possible changes in mutation rate without prior consideration of specific agents. The requirements for these two purposes are of course somewhat different. Studying the effects of exposure to a specific agent requires a control group comparable insofar as possible for all factors other than the variable agent. The two groups need not be representative of the total population so long as they are comparable to each other. On the other hand, monitoring for any change due to unidentified agents requires wide sampling in order to include

---

*Supported in part by research grant GM 09326 and contract 70-2288 from the National Institutes of Health.

persons exposed to the variety of potential mutagens available to us.

Conventionally, discussions of mammalian mutations have dealt with germinal mutations. The methodology and the definitions are much more clearly defined for germinal than for somatic mutations. The enormously greater sensitivity in detection of somatic mutations makes their study of interest even though the quantitative meaning of somatic mutation rates is still poorly defined. In the case of chromosomal changes, somatic mutation is an established approach, although it has been used only on small high-risk groups. The ability to detect risk to individuals rather than populations has made chromosomal studies of astronauts and other select persons routine. Not all mutagenic agents are effective in breaking chromosomes, and Shaw has discussed some of the problems of monitoring somatic chromosomes elsewhere in this volume. Alternate methods of observation are needed to assure response to the great variety of mutagens. In this discussion, attention will be directed toward possibilities for detecting non-chromosomal events--point mutations--in somatic cells.

## The Frequency of Point Mutations

First, let us consider the nature of the events which give rise to point mutations, leading in turn to structurally altered protein products. The function of the protein may or may not be altered, but the protein must be identifiable if a usable system is to be developed. Therefore, the detectable variants are limited essentially to amino acid substitutions. Small deletions such as occur with hemoglobin Freiburg and hemoglobin Gun Hill would also qualify, but these must be exceedingly rare compared to substitutions. Conventional mutation rate studies in man have been concerned almost exclusively with dominant traits where the nature of the gene change is completely unknown. In a recent summary of mutation studies assembled by Vogel (1970), eleven dominant mutations were listed, in none of which is the site of gene action known, much less the primary nature of the gene alteration. The conventional outlook on metabolic disorders is that the defective gene product has a virtual or total loss of activity. This has been fairly simple to document in the case of recessive metabolic disorders. It need not be true in the case of dominant disorders, however. Even if the dominant disorders which have been studied result from an allele giving rise to no activity, what does this mean in terms of other kinds of mutation rates? What proportion of all amino acid substitutions and indeed of small deletions result in total loss of activity? The answer could be 10 percent or 90 percent. The thinking in terms of mutation has generally been that the zero activity alleles represent the contributions of many different mutations within the structural locus. This may be true, but convincing arguments to the contrary can be made with existing data.

Assuming a nucleotide substitution rate of $10^{-8}$ per generation gives a rate of $10^{-5}$ for a molecule such as hemoglobin resulting from some 300 codons. Possibly a third of these mutations will lead to an amino acid substitution involving a change in charge. The detectable rate would then be on the order of $3 \times 10^{-6}$ mutations per locus (or molecule) per generation. Interestingly, this is the same order of magnitude that has been established on the basis of dominant phenotypes. It is also consistent with the very limited direct observations from England on hemoglobin mutations (cited in Vogel, 1970).

## Somatic Point Mutations

The monitoring of somatic mutations offers enormous advantages in theory over germinal mutations. Each person studied corresponds to an entire population studied for germinal mutation. With a small number of persons, specific exposure groups can be constituted, and dose-response curves can be constructed. The problem, of course, is that at present there is no effective way to measure the rate of somatic mutations.

Let us consider first the expected frequency of somatic mutations. Earlier $3 \times 10^{-6}$ was used as the rate of detectable variants of hemoglobin per generation. To simplify the calculations I shall relate the calculations to 100 codons, giving a rate of $1 \times 10^{-6}$ per generation. The rate per cell generation would be this figure divided by the number of cell generations. To add another simplifying assumption, I shall assume 50 cell generations, giving rise to approximately $10^{15}$ cells. This figure ignores differential growth rates among tissues and ignores the fact that, in a fully developed organism, some tissues cease cell division while others continue to divide rapidly.

If we divide $1 \times 10^{-6}$ by 50, we get $2 \times 10^{-8}$ as the mutation rate of electrophoretic protein variants per 100 codons per cell generation. The number of cell generations may be very large, at least in those tissues such as the blood-forming elements and the gut epithelium which undergo continual and rapid replication. There, the mutation rate per cell generation should be multiplied by the number of cell generations to get the expected number of cells mutant for a particular protein. By various assumptions one may reasonably expect one mutant cell in $10^4$ cells. The possibility of detecting mutant cells in this frequency is promising, even though any screening technique will respond only to a portion of the mutants.

Several theoretical complications may be imagined. One is the appropriateness of multiplying the mutation rate by the number of cell generations to obtain the total frequency of mutants. Some types of mutations almost certainly occur as a function of absolute time. Others very likely are functions predominantly of cell replication. Probably the best factor would be some complex function both of absolute time and replication.

The conclusion I should like to draw is that at present we have little basis for expecting one frequency of mutant cells versus another. The actual frequency of somatic point mutations can only be established empirically by the development of suitable test systems. The necessary methodology is not now available, but I believe that it can be developed by extension of existing techniques. The sensitivity of biochemical reactions is certainly adequate to establish the phenotypes of individual cells, and indeed this is routinely done in clinical histological and cytological laboratory studies. The difference here is that one is looking for a scarce phenotype rather than the mass phenotype.

I have described elsewhere (Sutton, 1971) an approach based on the observation that certain variants of glucose-6-phosphate dehydrogenase have broadened specificity. Whereas the normal enzyme will utilize only glucose-6-phosphate as substrate, two of the known human variants will also utilize 2-deoxyglucose-6-phosphate and galactose-6-phosphate at rates greater than the normal substrate. We have detected rare white blood cells able to utilize 2-deoxyglucose-6-phosphate. The frequency is on the order of one per thousand. I am unwilling as yet to call these variant cells mutants. A suitable proof of transmissibility must be devised which will demonstrate what portion if any of these variants truly are mutants. Whether or not this particular system proves to be useful in the detection of somatic mutations, the principle on which it is based is valid and other systems can be derived which will work. The greatly increased discrimination of systems for measuring somatic mutation would appear to justify a substantial effort in this direction.

Other potential systems for detecting somatic point mutations include that of Albertini and DeMars (1970) in which mutants for hypoxanthine-guanine phosphoribosyl transferase can be detected. In their system, fibroblasts which are deficient in this enzyme can be selected in culture, and these authors have isolated a number of such mutants. On the one hand, the system would appear promising as an *in vitro* test system for mutagenicity of specific substances. The other possibility is that the isolation procedure might be useful in detecting pre-existing mutants from freshly isolated tissue. If such systems could be developed for recognizing pre-existing mutants among lymphocytes, with high efficiency of cloning, this should provide a very important means of assessing the mutational experience of individuals.

### Repair of Somatic Mutations

It has become apparent that a variety of processes exist in nuclei for repair of mutation (reviewed in Fishbein, Flamm, and Falk, 1970). The studies of repair of ultraviolet damage to DNA have been especially successful, and the demonstration of a block in this repair system in the human disease xeroderma pigmentosum (Cleaver, 1969) has made the concern with repair mechanisms clearly relevant to man. The techniques for study of ultraviolet repair in human cell culture have been established. The incorporation of labeled thymidine in place of excised thymine is a measure of the amount of repair and hence, indirectly, the amount of damage.

It should be possible ultimately to develop methods sensitive enough to detect repair of DNA alterations that occur *in vivo* in man. With the exception of the one mechanism for repair of thymine dimers, little is known of human repair systems. The potential utility of such systems is sufficiently great to justify a special effort in their study. Monitoring of current repair would give the fastest possible indication of potential genetic damage. The relationship of repair to permanent damage would likely be complex, but this too is susceptible to study.

## Sentinel Phenotypes for Somatic Mutation

Discussions of the use of sentinel phenotypes have been primarily with respect to detection of germinal mutation. In that case the genetics is clean, but it is generally agreed that such investigations have severe limitations. The use of sentinel phenotypes for monitoring risk of somatic mutations has not been used directly. Rather, high risk populations have been monitored for development of sentinel phenotypes. I refer here to development of malignancy in persons exposed to radiation, such as radium workers, uranium miners, and survivors of atomic bombs. A difficulty has been the long latency period between exposure and development of overt disease. Also, the origin of these malignancies has not been shown unequivocally to be due to mutation. This latter may not be too important an issue, since mutagenicity and carcinogenicity so often go hand-in-hand. Further, none of us would question the value of knowing the risk due to environmental carcinogens, whatever the basic mechanisms of carcinogenesis.

Recent studies by Knudson (1971) on retinoblastoma give hope that this and possibly other diseases might be used to assess somatic mutation rates. It has long been known that many cases of retinoblastoma are associated with dominant inheritance. A person with the dominant allele has a very high risk of developing one or more tumors, although a few such persons may escape. Sporadic cases also occur and are most often unilateral and not transmitted to offspring. Knudson has proposed, and the detailed observations agree, that a tumor results when two mutant genes occur together in the same cell. One mutant allele may be transmitted in a Mendelian fashion, requiring only one additional somatic mutation to give rise to a tumor. Otherwise, two somatic mutations must occur in order to produce a tumor.

An effective system for recording the occurrence of retinoblastoma, leukemia, and other malignancies might be an important part of a program to monitor mutational risk. Epidemiologists should be supported in their efforts to establish risks and relate the risks to environmental variables. They should be encouraged to give special attention to those diseases in which a mutational component is implicated.

## Conclusion

Although the case for monitoring somatic mutation is easy to make, the means to do so are not available. Some promising directions have been identified, but the list is not exhaustive. Research into the methods for detecting somatic mutations has been limited. With the concern that the changing environment may be greatly increasing the mutational risk, the need to develop such methods is clear. Ultimately, there is no alternative to looking at man himself if we are to know what is happening to him.

## References

Albertini, R. J., and R. DeMars (1970). *Science*, *169*, 482.

Cleaver, J. E. (1969). *Proc. Nat. Acad. Sci. (U.S.)*, *63*, 428.

Fishbein, L., G. Flamm, and H. Falk (1970). *Chemical Mutagens*. New York, Academic Press.

Knudson, A. G., Jr. (1971). *Proc. Nat. Acad. Sci. (U.S.)*, *68*, 820.

Sutton, H. E. (1972). In *Monitoring, Birth Defects, and Environment--The Problem of Surveillance*. New York, Academic Press.

Vogel, F. (1970). In *Chemical Mutagenesis in Mammals and Man*, F. Vogel and G. Rohrborn, editors. Berlin-Heidelberg, Springer-Verlag.

# PESTICIDAL, INDUSTRIAL, FOOD ADDITIVE, AND DRUG MUTAGENS

*Lawrence Fishbein*

## Introduction

Man is assaulted wittingly and unwittingly by a broad spectrum of environmental insults that are both synthetic and naturally occurring. A number of these toxicants are carcinogenic, teratogenic, and/or mutagenic. It is especially this latter category that has the greatest portent for possible future genetic disaster.

The prime objective of this review of pesticidal, industrial, food additive, and drug mutagens is to put into basically an environmental perspective or framework the above four major areas of human exposure to these chemical mutagens. Hence, areas of consideration of the chemical mutagens that are germane include: their occurrence, the scope of usage (where known), the chemical form(s) (original, degradation, or metabolic) of encounter; and portals of entry (e.g., ingestion as residues in food and water, atmospheric pollutants, or industrial exposure).

It is convenient to arrange the chemical mutagens in terms of their utility, e.g., pesticides, industrials, food additives, and drugs, as well as their basic chemical structural similarities or mode of action, e.g., alkylating agents, although it is recognized that there are obvious overlaps, e.g., alkylating agents that have pesticidal, industrial, and drug utility. Sections I, II, III, and IV which follow illustrate the trade or common names, chemical names, and the types of mutations reported in the literature. Literature references to the point mutations and chromosomal aberrations of the listed agents are outlined in the compendiums of Fishbein, Flamm, and Falk (1970) and Barthelmess (1971). Shaw (1970) has recently reviewed the subject of human chromosome damage by chemical agents.

The pesticidal mutagenic agents constitute a prime area of human concern. Included in this category are fumigants, herbicides, fungicides, seed sterilants, insecticides, and chemosterilants. A number of these agents are of major importance and are used in considerable amounts in such forms as granular formulations, dusts, powders, sprays, foams, and aerosols, and hence man can be exposed to these agents via consumption of toxicant residues in food, handling of the agents per se, or through ecological distribution.

All organic pesticides, to a varying degree, are metabolized in living organisms and/or are degraded environmentally (e.g., photolytically, thermally). The extent and nature of these transformations vary with the agent causing them and with pesticidal chemical structure and time being important factors, e.g., the transformation of some of these agents occurs in a matter of minutes, while that of others requires months or even years. The chemical reactions involved include hydrolysis, oxidation, reduction, dehalogenation, desulfurization, ring opening, isomerization, and conjugation.

These considerations are of fundamental importance in any serious mutagenic evaluation. It is imperative to realize that although it may be experimentally convenient to assess the purified chemical agent in question in diverse mutagenic test systems, cognizance must be made of the myriad possibilities of man absorbing and ingesting a host of related and structurally unrelated entities arising from not only the metabolism and environmental degradation of the agent, but via precursors and degradation products attendant with the synthesis of the agent. This aspect arises because of both economic necessity and the possibilities that the reaction impurities possess sufficient pesticidal activity per se to warrant their inclusion or that their physical and chemical properties argue against their facile removal.

Both industrial and drug mutagens represent categories containing a rich variety of organic structures and range of application. The largest number of mutagens in these categories are the alkylating agents, which can be further subdivided on the basis of their functional groups, e.g., aziridines, mustards, epoxides, lactones, aldehydes, nitrosamines, and dialkyl sulfates. It is important to note their reactivity with DNA either by an SN 1 or SN 2 mechanism (both appearing to react preferentially with the N-7 of guanine). The reactivity of this class of agents (from a consideration of the myriad byproducts in their preparation, types of residues, and cellular interactions) is also stressed. Other categories of agents that warrant attention as drug mutagens are a variety of antibiotics, acridines, and folic acid antagonists that possess a broad spectrum of utility, e.g., antineoplastics, antimalarials, antipyretics, and sedatives. Although the synthetic food and feed additive mutagens are comparatively few in number and are relatively simple in structure, they are obviously of major importance as a prime source of human contact. Added to this burden are the naturally occurring mutagenic agents such as cycasin, a number of pyrrolizidine alkaloids, bracken fern, aflatoxin $B_1$, and patulin. The polynuclear hydrocarbons 3,4-benzpyrene and 1,2,5,6-dibenzanthracene are also included in this category because of their occurrence in trace amounts in some foods as well as components in air pollution.

To add additional dimensions to the overall problem of man's exposure to the spectrum of environmental mutagens are the instances of mutagens used as synthetic precursors in the preparation of a number of important agents in the categories discussed above. It is also important to stress the fact that not only are chemical mutagens (both synthetic and naturally occurring, as well as their metabolic and/or degradation products) capable of producing one or more of a variety of acute and chronic toxicities, but they may also interact _in vivo_ to produce synergistic or potentiating effects.

# SECTION I. PESTICIDAL MUTAGENS
## Herbicides, Fungicides, Insecticides, Chemosterilants, Fumigants, and Seed Sterilants

1. Maleic Hydrazide

   1,2-dihydro-3,6-pyridazine-dione

Maleic hydrazide is prepared commercially via the reaction of hydrazine with maleic anhydride and has extensive application as a herbicide, fungicide, growth inhibitor, and growth regulator. Approximately 1.4 million pounds of maleic hydrazide are used annually. Its uses include the prevention of sucker production in tobacco plants; growth control of weeds, grass, and foliage; inhibition of sprouting of potatoes, onions, and stored root crops; and the protection of citrus seedlings against frost damage. Maleic hydrazide is carcinogenic and has exhibited mutagenicity in *Drosophila*; mitotic inhibition and chromosomal aberrations in *Vicia faba*, *Allium*, barley, maize, mitotic tissues of oats, corn, and soybeans; and cytochemical effects on cultured mammalian cells.

2. Captan

   N-(trichloromethylthio)-4-cyclohexene-1,2-dicarboximide

Captan is a general fungicide used for the treatment of folia and soil and against seedborne diseases including apple scab, grape mildews, corn seed infections, and many fruit, vegetable, and ornamental plant diseases. Captan is teratogenic to the chick embryo and mutagenic in *E. coli* and cells cultured *in vitro*.

3. DDT

1,1,1-trichloro-2,2-di-(4-chlorophenyl)-ethane

Cl–C₆H₄–CH(CCl₃)–C₆H₄–Cl

DDT is the best known, cheapest, and most effective of the synthetic insecticides. It is prepared in a great variety of formulations both for home and agricultural usage. The agricultural products are generally in the form of wettable powders and sprays containing 1-10% and 25-50%, respectively, of DDT. For home preparations against flies, mosquitoes, roaches, bedbugs, etc., it is commonly marketed in combination with the synergist piperonyl butoxide and a pyrethrin for rapid breakdown. The amounts of DDT in use in the United States have dwindled slowly so that approximately 51 million pounds of DDT were used in 1964 (although 141 million pounds were produced). It is made via the condensation of chloral with monochlorobenzene in sulfuric acid, producing a commercial product that illustrates, almost to the extreme, the numbers of constituents that can be present in a commercial pesticide. Besides containing the major product, p,p'-DDT (approximately 70%), technical DDT contains other isomers, degradation products, and reaction byproducts, e.g., o,p'-DDT, p,p'-DDD, o,p'-DDD, bis (p-chlorophenyl)-sulfone, chlorobenzene, p-dichlorobenzene, 2-trichloro-1-p-chlorophenyl ethanol, 2-tris-chloro-1-o-chlorophenyl-ethyl-p-chlorobenzene sulfonate, o-chloro-α-p-chlorophenyl acetamide, and at least half a dozen other components. It should also be noted that some of the byproducts are active insecticides, though none of them are as toxic as p,p'-DDT. The extremely low water solubility, as well as relative environmental stability to ultraviolet and thermal parameters, has led to the remarkable persistence of DDT (the biological half-life is estimated to be in excess of 10 years). Hence, traces of it or its degradation products, their ubiquitous ecological distribution, long-lasting effects upon non-target ecosystems, and storage in mammals, coupled with the recent findings of hepatomas in rats on chronic feeding of DDT, have caused a serious review of the continued use of DDT. Although its use has recently been severely curtailed in the United States, it is most probable that it will be generously employed in Africa, Asia, India, and Latin America, primarily for the continued eradication of malaria and other vector diseases. Legator and colleagues have recently reported the mutagenicity of DDT in rats (dominant lethal test) and in human cells _in vitro_. This finding alone should cause a serious re-evaluation of this agent.

4. Aramite

2-(p-tert-butylphenoxy)-1-methylethyl-2-chloroethyl sulfite

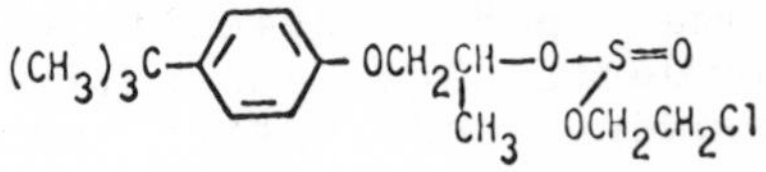

$(CH_3)_3C-C_6H_4-OCH_2CH(CH_3)-O-S(=O)-OCH_2CH_2Cl$

Aramite is a non-systemic acaricide used on nursery stock, shade trees, cotton, and some fruit trees. It has been shown to produce liver and bile duct tumors in rats when incorporated into the diet, as well as producing a significant incidence of bile duct and gall bladder tumors in dogs when incorporated into their diet (containing 828 ppm of aramite over a 3-1/2 year period). The carcinogenicity of aramite has been attributed to its alkylating properties attendant with the chloroethyl moiety in its structure. Aramite is mutagenic in _Drosophila melanogaster_.

5. DDVP

Vapona
2,2-dichlorovinyl-dimethyl phosphate

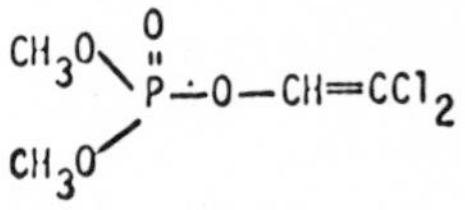

$(CH_3O)_2P(=O)-O-CH=CCl_2$

DDVP is prepared via the interaction of two mutagenic agents, trimethyl phosphate and chloral, and is employed as an insecticide, fumigant, and in veterinary medicine as a helminthic. It is widely used as the active compound in household vaporizing resin strips (in certain pest strips as a 20% solid solution in plastic). DDVP causes chromosome aberrations in onion root tip cells and _Vicia faba_ and increased mutation rate in a streptomycin-dependent Sd-4-mutant of _E. coli_ B. The alkylation of calf thymus DNA has recently been demonstrated. Since many organophosphorus insecticides are of the triester type and are likely to be alkylating agents (as DDVP), the investigation of their mutagenicity in appropriate test systems has been advocated in many quarters.

## 6. Aziridines

Tepa
APO, Aphoxide
Tris(1-aziridinyl)-phosphine oxide

$H_2C$ — $CH_2$
N
$H_2C$ N—P=O
$H_2C$
N
$H_2C$ — $CH_2$

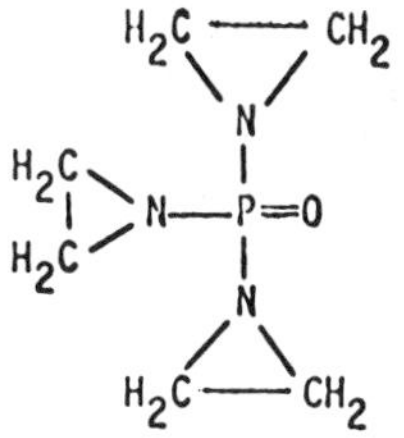

Metepa
Mapo, Methaphoxide
Tris-1-(2-methylaziridinyl)-phosphine oxide

$CH_3$
$H_2C$ — C
H
N
$H_2C$
N—P=O
H-C
N
$H_3C$
$H_2C$ — C—$CH_3$
H

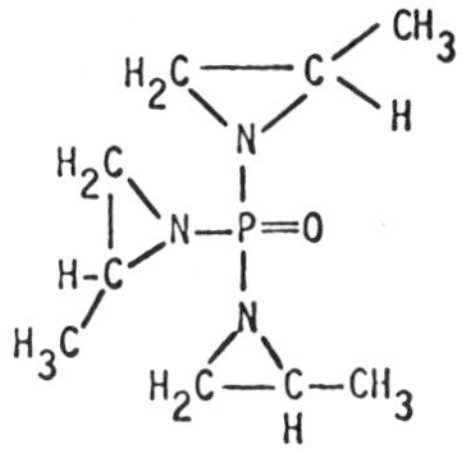

Apholate
2,2,4,4,6,6-hexakis(1-aziridinyl)-2,2,4,4,6,6-hexahydro-1,3,5,2,4,6-tri-aza-triphosphorine

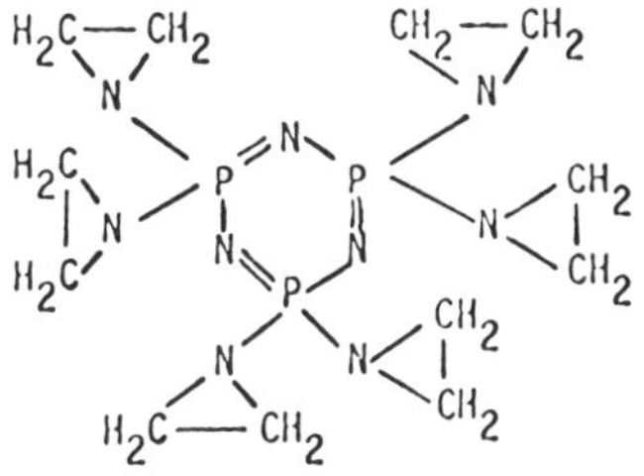

$H_2C$ — $CH_2$
N
$CH_2$ — $CH_2$
N
$H_2C$
N
$H_2C$
P=N P
N P N
N
$CH_2$
$CH_2$
N
N
$CH_2$
$CH_2$
$H_2C$ — $CH_2$

Thiotepa
Tris(1-aziridinyl)-phosphine sulfide

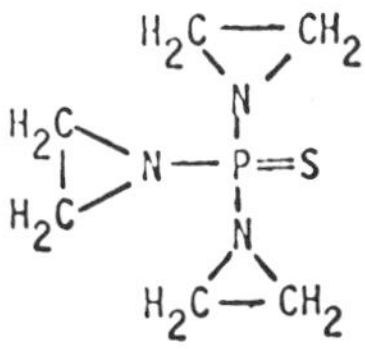

$H_2C$ — $CH_2$
N
$H_2C$
N—P=S
$H_2C$
N
$H_2C$ — $CH_2$

Tepa has been the most extensively employed of all the aziridinyl phosphine oxides. Its industrial applications include the flame-proofing of textiles; preparation of water-repellant, wash and wear, crease-resistant fabrics; treatment of paper and wood; and use in dyeing, printing, adhesives, and binders. The utility of tepa (as well as other aziridines) is illustrative of the ability of these compounds to act as cross-linking agents for polymers containing active hydrogen groups such as carboxyl, phenol, sulfhydryl, amide, and hydroxyl. Interest in tepa and related compounds has been intensified in recent years by the discovery of their activity as insect chemosterilants (anti-fertility agents) and their potential utility as new and powerful tools for insect control and eradication. However, inherent in such programs of eradication is the dispersion of large numbers of treated insects into the environment. Hence, there is great concern in elaborating how much active chemosterilant remains on and in the insect that is to be released and how long it will persist. The residues and persistence of chemosterilants (e.g., tepa, metepa, apholate, and thiotepa) as well as their various metabolic and degradation products in a number of insect species have been well documented. To date, the chemosterilants have proven effective in insect control (e.g., houseflies, weevils, gypsy moths, Japanese beetles, Mexican fruit flies, and codling moths) in selected experimental areas, and it appears that their general applications will have to be rigidly controlled. Tepa is mutagenic in the parasitic wasp, *Habrobracon*, in mice (dominant lethal test), *Neurospora*, *E. coli*, and bacteriophage T4. Metepa is mutagenic in mice (dominant lethal test), *Neurospora*, and bacteriophage T4, and induces chromosome aberrations in *Vicia faba*. Metepa is also teratogenic in the rat. Apholate induces mutations in *Neurospora*, dominant lethal mutations in mature sperm, and gonial cell death in the housefly *Musca domestica*. Thiotepa is mutagenic in mice (dominant lethal test), screw worm fly *Cochliomyia hominivorax*, and bone marrow cells of mice, and induces chromosome aberrations in human chromosomes and bone marrow cells of mice. Thiotepa has demonstrated utility in the temporary palliation of certain cancers, e.g., breast cancer, Wilm's tumor, chronic lymphatic leukemia, ovarian carcinoma, and Hodgkin's disease.

7. TEM

Triethylenemelamine

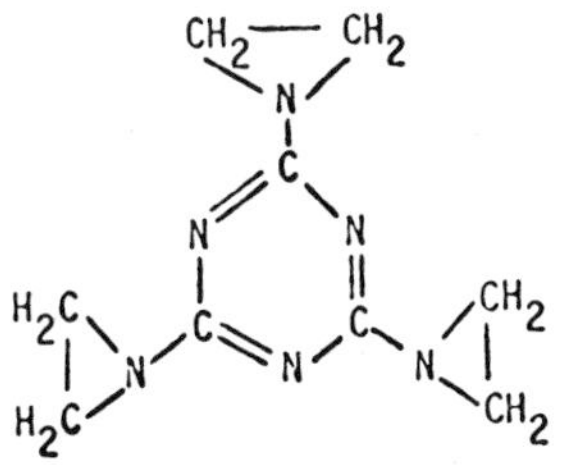

Triethylenemelamine, prepared from ethylenimine and cyanuric chloride, is used in the manufacture of resinous products, as a cross-linking agent in textile technology, and in the finishing of rayon fabrics and waterproofing of cellophane. Major interest, however, has been related to its medical utility as an antineoplastic agent and as a chemosterilant for the housefly *Musca domestica*, screwworm, and oriental, melon, and Mediterranean fruit flies. The mutagenic activity of TEM has been demonstrated in mice, *Drosophila*, *Musca domestica*, *Neurospora*, and the screwworm fly *Cochliomyia hominivorax*, and it induces chromosome aberrations in mice, *Drosophila*, cultured human leukocytes, barley, *Vicia faba*, *Allium cepa*, *S. typhimurium*, and *E. coli*.

8. Hemel
2,4,6-tris-dimethyl-amino-1-triazine

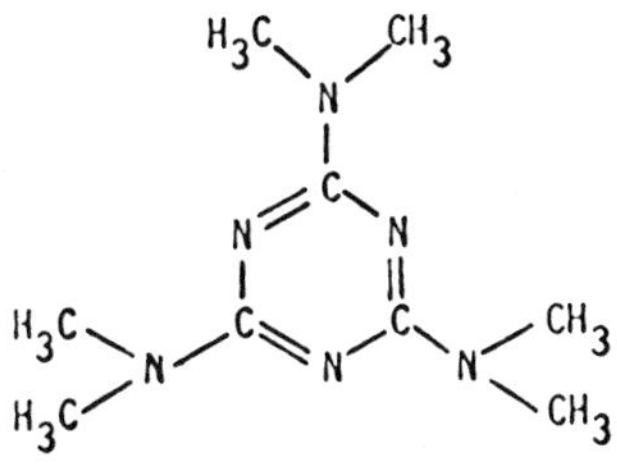

Hempa
Hexamethyl phosphoric triamide

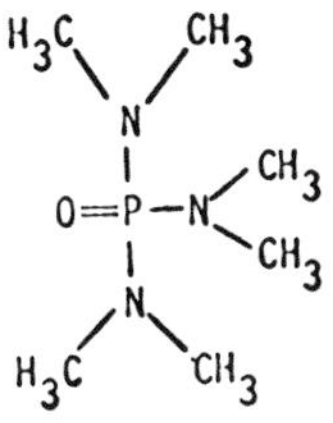

Hemel and hempa are thermally stable non-alkylating analogs of tretamine and tepa, respectively, and are effective chemosterilants for houseflies, Musca domestica. Hempa induces a high frequency of recessive lethal mutations in the sperm of the wasp, Bracon hebetor, and a marked anti-spermatogenic effect in rats and mice. Hemel is mutagenic in Musca domestica.

9. Organomercurials

Ethylmercuric chloride $C_2H_5—Hg—Cl$

Phenylmercuric salts $C_6H_5—HgX$
X = OH, $NO_3$; $OCOCH_3$

Methylmercuric salts $CH_3—Hg—X$
X = OH; dicyandiamide

Methoxyethylmercury chloride $CH_3OCH_2CH_2HgCl$

A number of organomercurials, e.g., phenyl-, ethyl-, and methylmercuric salts, have enjoyed (until recently) wide utility as fungicides and seed disinfectants. Phenylmercuric acetate (PMA) is a powerful eradicant foliage fungicide with industrial applications for fungal and bacterial control as well (U. S. production of PMA in 1966 was 502,000 pounds). Ethyl- and methylmercuric salts are used for the seed treatment of cotton, flax, grains (rice, wheat, barley, rye, corn), sugar, pea, millet, and certain vegetable seeds. The genetic effects of the organomercurials (Fishbein, 1971) include the mutagenic activity of Mercuran (fungicide containing 2% ethylmercuric chloride and 12% hexachlorocyclohexane) in germinating apple seeds; somatic mutations produced by phenylmercuric hydroxide and phenylmercuric nitrate in flowering plants (seedlings of Raphanus and Zea) and induction of polyploid nuclei; sticky chromosome and chromosome fragments in root tips of Allium cepa; cytological effects on root cells of Allium cepa of methylmercury dicyandiamide, methylmercuric hydroxide, phenylmercuric hydroxide, methoxyethylmercuric chloride, and the fungicide Panogen (containing methylmercury dicyandiamide as the active component); cytological effects of inorganic, phenyl-, and alkylmercuric compounds (e.g., phenyl-, ethyl-, and butylmercuric chloride) on HeLa cells; histological and cytological effects of ethylmercuric phosphate in corn seedlings; the C-mitotic action of "Granosan" (fungicide containing ethylmercuric chloride) and Agrimax M (containing ethylmercuric chloride and phenylmercury dinaphthyl methanedisulfonate, respectively); the genetic effects of methylmercuric hydroxide, phenylmercury acetate and methoxyethylmercury chloride in *Drosophila melanogaster*; and the induction of chromosome breakage in humans with methylmercury. Merthiolate (sodium ethylmercury thiosalicylate), which is generally used as an antiseptic and also has application as a fungicide for cotton seed treatment, is mutagenic in *Drosophila melanogaster*. It has also recently been reported that a number of individuals regularly consuming large amounts of fish with elevated methylmercury concentrations have significantly higher frequencies of lymphocyte chromosome aberrations than comparable control persons. The complexing and denaturation of DNA by methylmercuric hydroxide has also been reported. The teratogenicity of phenylmercuric acetate in mice, the embryotoxic effects of "methylmercury" in mice, and in humans the intrauterine effects of methylmercury dicyandiamide in Sweden and "methylmercury" in Japan have also been described.

## 10. Epoxides

Ethylene oxide — $CH_2$—$CH_2$ (ring closed through O)

Propylene oxide — $H_2C$—$CHCH_3$ (ring closed through O)

Ethylene oxide is produced on an enormous industrial scale (approximately 2.5 billion pounds per year) by the action of alkali on ethylene chlorohydrin (a mutagen) or by catalytic oxidation of ethylene in air. Epoxides such as ethylene oxide and propylene oxide owe their industrial importance to their high reactivity, which is due to the ease of opening of the highly strained three-membered ring. A host of industrially important chemicals are synthesized from ethylene oxide, e.g., ethylene glycol, diethylene glycol, dioxane, carbowax, methyl carbital, ethylene chlorohydrin, monoethanolamine, acrylonitrile, and surface active agents. Ethylene oxide also has utility as a solvent and plasticizer in combination with other chemicals and in the production of high-energy fuels, diverse plastics, textile auxiliaries, and hydroxyethylated cellulosis fibers and starch. The area of utility which is most important in terms of residues of toxic and mutagenic agents is that of gas sterilization and fumigation. All of the compounds most actively employed in gaseous sterilization, e.g., ethylene oxide, propylene oxide, β-propiolactone, formaldehyde, and methyl bromide, are alkylating agents. (The mode of action upon microbes appears to be a non-specific alkylation of such chemical groups as -OH, -NH-, and -SH, with loss of a hydrogen atom and the production of an alkyl hydroxyethyl group.) The use of ethylene oxide for sterilization has thus raised a number of significant questions regarding (a) the possible entrapment and/or interaction of ethylene oxide in plastic, pharmaceuticals, and food, which may then exert a toxic effect when placed in contact with living tissue, and (b) the effect of sorbed ethylene oxide on the possible changes in the physical and chemical properties of the medical plastics per se. The literature (Fishbein, 1969) is replete with examples of not only trace amounts of ethylene oxide entrapped for extended periods of time in plastic devices, drugs, and foods undergoing sterilization and/or fumigation, but also chemical interactions of this alkylating agent with a number of the reactants as well, e.g., antibiotics, steroids, plasticizers, and vulcanization accelerators of plastics and rubber devices. Residual ethylene oxide might produce a toxicity from the following sequences of oxidation: epoxide → glycol → glyoxal → glyoxalic acid → glycolic acid → oxalic acid. The area of greatest concern is both in residual epoxides and in the formation of toxicants resulting in the treatment of foods per se with ethylene oxide (as well as propylene oxide). The primary residues formed are glycols and chlorohydrins. Chlorohydrins in foodstuffs that are insufficiently involatile to be persistent under food processing conditions have indeed been demonstrated. (It is important to note that ethylene chlorohydrin has recently been shown to be mutagenic in _Klebsiella pneumoniae_.) Other areas of utility of ethylene oxide include its use in the tobacco industry to shorten the aging process and reduce the nicotine content in tobacco leaves. Ethylene oxide has also been found in cigarette smoke from tobacco treated with oxyethylene docosanol, as well as non-treated tobacco, and also in the smoke of cigarettes that possess charcoal filters. Ethylene oxide is mutagenic in *Drosophila*, *Neurospora*, and barley, and induces chromosome aberrations in maize, barley, and *Vicia faba*.

## 11. Miscellaneous Potential Pesticidal Mutagens

A number of pesticides warrant special consideration (and future mutagenic evaluation) because of structural and biological similarities with known mutagens as well as their known effects (in some instances) on plants. For example, a number of pesticides have been shown to produce mutations to a varying degree in barley treated at 1000 ppm levels for 12 hours (and their relative efficiency compared to 5,500 R of x-rays). A number of the more active and widely used agents are tabulated below in terms of trade name, chemical name, and utility, in decreasing order of relative mutagenic efficiency.

| | | |
|---|---|---|
| Linuron | 3-(3,4-dichlorophenyl)-1-methoxy-1-methyl urea | herbicide |
| Simazine | 2-chloro-4,6-bis(ethylamino)-1,3,5-triazine | herbicide |
| Atrazine | 2-chloro-4-ethylamino-6-isopropylamino-s-triazine | herbicide |
| Monuron | 3-(p-chlorophenyl)-1,1-dimethyl urea | herbicide |
| Embutox (2,4-DB) | 4-(2,4-dichlorophenoxy) butyric acid | herbicide |
| Sevin (carbaryl) | 1-naphthyl-N-methyl carbamate | insecticide |
| Dicamba | 3,6-dichloro-2-methoxybenzoic acid | herbicide |

In addition, there are a number of pesticides having chemical structures that are known to affect DNA via alkylation and hence warranting further evaluation for potential mutagenicity. These include the insecticidal epoxides such as dieldrin and endrin and the fumigants methyl bromide and ethylene dibromide. Other agents that warrant further scrutiny include (a) the radical-producing agents which can induce large chromosome alterations but not point mutations (e.g., bipyrridylium quarternary salts, diquat and paraquat), and (b) carbamates and thiocarbamates such as barban, mobam, temik, vapam, thiram, and ziram, which do not affect DNA directly, but their enzymic products such as N-hydroxycarbamates (and other intermediates) produce radicals resulting in inactivating DNA alterations. Hence, chromosome breaks and large chromosome alterations, but not point mutations, can be induced.

## SECTION II. INDUSTRIAL MUTAGENS

### 1. Formaldehyde HCHO

Formaldehyde is used in enormous quantities (e.g., 3.4 billion pounds in U. S. in 1965) generally in the form of aqueous solutions 37% to 50% formaldehyde by weight. The major uses of formaldehyde and its polymers (the hydrated linear polymer paraformaldehyde and the cyclic trimer S-trioxane) are in the synthetic resin industry (e.g., in the production of thermosetting and oil-soluble resins and adhesives which account for 50% of the total production). Large quantities are used in the manufacture of a broad spectrum of textiles, paper, fertilizer, miscellaneous products, and specialty chemicals. The agricultural uses of formaldehyde include disinfection of seeds; prevention of scab in potato, wheat, barley, and oats; and fungicidal and preservative applications. In the textile industry, formaldehyde alone and in the form of its N-methylol derivatives are extensively employed for the production of crease-proof, crushproof, flame resistant, and shrinkproof fabrics. Formaldehyde has been found widely in man's environment, e.g., in tobacco leaf, tobacco smoke, incinerator effluents, automobile and diesel exhaust, and as residues from treated textiles (as described above). Formaldehyde is extremely reactive and will react with practically every type of organic chemical, primarily leading to the formation of methylol or methylene derivatives. The mutagenicity of formaldehyde has been described most extensively for *Drosophila* and established for *Neurospora* and *E. coli*.

2. Acetaldehyde $CH_3CHO$

The amount of acetaldehyde produced from ethanol (one of three major synthetic routes) far exceeds 900 million pounds yearly in the United States. Acetaldehyde is an intermediate in the manufacture of a host of important products including acetic acid, acetic anhydride, butyl alcohol, butylaldehyde, peroxy acetic acid, acrylonitrile, cellulose acetate, chloral, and vinyl acetate resins. It has been used as a preservative for fruit and fish, as a denaturant for ethanol, to prevent mold growth on leather, and as a solvent in the rubber, tanning, and paper industries. Acetaldehyde is the product of most hydrocarbon oxidants, occurs in traces in all ripe fruits, may form in wine and other alcoholic beverages exposed to air, and has been reported in fresh leaf tobacco, as well as in tobacco smoke and in automobile and diesel exhaust. Acetaldehyde is mutagenic in *Drosophila*.

3. Acrolein $CH_2{=}CHCHO$

Acrolein is prepared on a commercial scale by cross-condensation of acetaldehyde with formaldehyde using lithium phosphate or activated alumina as catalyst. It is largely used in the manufacture of the intermediate acrolein dimer 3,4-dihydro-2-formyl-2H pyran, which in turn is valuable as a starting point for the synthesis of a variety of chemicals (e.g., hexanetriol, hydroxy adipaldehyde, and glutaldehyde), and is useful in textile finishing, paper treating, and the manufacture of rubber chemicals, pharmaceuticals, plasticizers, and synthetic resins. Other major products of acrolein synthesis include methionine (used in supplementing food, and swine and ruminant feeds), acrylonitrile, pentaerythritol, and glycidaldehyde. Acrolein has been identified as an air pollutant in automobile smog and the atmosphere of paint and varnish plants and in both tobacco leaf and tobacco smoke. It is produced during the pyrolysis of humecants in tobacco such as glycerol and acrolein (as well as formaldehyde, hydrocarbons, organic peroxides, formic acid, sulfur dioxide, ammonia, and nitric oxide) and has been identified as a volatile contaminant in smog. Acrolein is mutagenic in Drosophila.

4. Ketene $CH_2{=}C{=}O$

The major uses of ketene are in the manufacture of acetic anhydride and the dimerization to diketene (an important intermediate in the preparation of acetoacetic esters, dihydroacetic acid, and cellulose esters used in the manufacture of fine chemicals, drugs, and insecticides. Ketene has utility as a rodenticide and is a very useful acetylating agent for ROH and $RNH_2$ groups and hence has broad organic synthetic utility. It is formed by pyrolysis of virtually any compound containing an acetyl group or by the reaction of ozone with olefins such as propylene, as well as by the photooxidation of hydrocarbons (olefins). Ketene is mutagenic in Drosophila.

5. β-Propiolactone

BPL

$$\begin{array}{ccc} H_2C & - & CH_2 \\ | & & | \\ O & - & C{=}O \end{array}$$

β-Propiolactone is produced commercially from formaldehyde and ketene in the presence of a catalyst such as zinc chloride. The common commercial product contains about 18% impurities, e.g., acrylic acid, acetic anhydride, and polymers. It possesses a broad spectrum of current and suggested industrial uses, e.g., wood processing, impregnation of textiles, protective coatings, intermediate in the preparation of insecticides, plasticizers, and medicinals, additive for leaded gasolines, and in the modification of tobacco flavors. BPL has been widely used as a sterilizing agent and as a solvent. Chemically, BPL is highly reactive due to the strained four-membered ring, is an alkylating carcinogen, is mutagenic in Vicia faba, Neurospora,

*E. coli*, and *Serratia marcescens*, and induces chromosome aberrations in *Vicia faba*, *Allium*, and *Neurospora*.

6. Glycidol

   2,3-epoxy-1-propanol

$$H_2C\!-\!CH\!-\!CH_2OH \quad (\text{C–C bridged by O})$$

Glycidol is used for the preparation of glycerol and glycidyl esters, ethers, and amines, which have utility as pharmaceutical intermediates and in water repellant textile finishings. Glycidol has been used as an antibacterial and antimycotic agent for food products. It is mutagenic in *Drosophila*, *Neurospora*, and barley, as well as inducing chromosome breakage in tissue cultures of mouse embryonic skin and Crocker mouse Sarcoma 180.

7. Epichlorohydrin

   1-chloro-2,3-epoxy-propane

$$H_2C\!-\!CH\!-\!CH_2Cl \quad (\text{C–C bridged by O})$$

Epichlorohydrin is employed extensively as a solvent for natural and synthetic resins, gums, cellulose esters and ethers, paints, varnishes, nail enamels, and lacquers, and as a raw material for the manufacture of glycerol and glycidol derivatives of epoxy resins. It is also used as a cross-linking agent in the crease-proofing of textiles, paper processing, waterproofing of materials, and as a curing agent for aminoplast resins. Epichlorohydrin is mutagenic in *Drosophila*, *Neurospora*, *E. coli*, and barley.

8. Di(2-3-epoxypropyl)-ether

$$H_2C\text{-}CH\text{-}CH_2\text{-}O\text{-}CH_2\text{-}CH\text{-}CH_2 \quad (\text{each terminal C–C bridged by O})$$

Di(2-3-epoxypropyl)ether is used in the preparation of trioxane copolymers, thermoset resins, vulcanizable polyethers and acetals, anion exchangers, polymers or flocculating agents, and as a diluent in aromatic amine-cured epoxy adhesives. It is mutagenic in bacteriophage T2 and *Neurospora* and induces chromosome breakage in *Vicia faba*, root tips of broad bean, and *Tradescantia*.

9. 1,2:3,4-Diepoxybutane

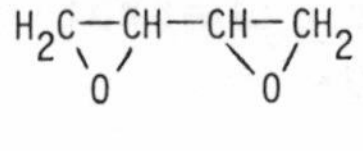

DEB
Butadiene diepoxide

$H_2C—CH—CH—CH_2$ (each terminal pair bridged by O)

1,2:3,4-diepoxybutane is used in the prevention of microbial spoilage, curing of polymers, as a cross-linking agent for textile fibers, and as an intermediate in the preparation of erythritol and pharmaceuticals. DEB is mutagenic in *Drosophila*, *Neurospora*, *E. coli*, *Salmonella typhimurium*, *Saccharomyces cerevisiae*, barley, maize, tomato, *Vicia faba*, *Penicillium*, *Arapidopsis thaliana*, housefly *Musca domestica L.*, and mammals, in addition to inducing chromosome aberrations in *Allium cepa*, *Vicia faba*, *Drosophila*, mammals, and mouse lymphocytes in vitro.

10. Dimethysulfate $CH_3OSO_2OCH_3$
DMS

Diethylsulfate $C_2H_5OSO_2OCH_3$
DES

Dimethyl- and diethylsulfates are extensively employed as alkylating agents both in the laboratory and in industry. The utility of DMS includes methylation of cellulose, preparation of alkyl ethers of starch, and solvent for the extraction of aromatic hydrocarbons. DES is used in the preparation of alkyl lead compounds, stabilization of organophosphorus insecticides, etherification of starch, finishing of cellulosis yarns, and as a catalyst in olefin polymerization. DMS is mutagenic in *Drosophila*, *E. coli*, and *Neurospora*, and induces chromosome breakage in plant material. DES is mutagenic in *Drosophila*, *E. coli*, bacteriophage T2, *Aspergillus nidulans*, peas, wheat, barley, *Neurospora*, and *S. pombe*.

11. 1,3-Propanesultone

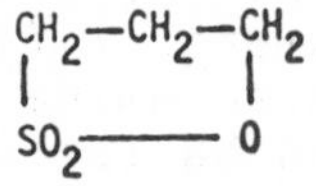

$CH_2—CH_2—CH_2$ / $SO_2$——O (ring)

Propanesultone represents a new class of bifunctional alkylating agents which react hydrolytically with a wide variety of nucleophilic substances to introduce the propylsulphonic function and which, because of their reactive nature, serve as useful intermediates in the preparation of detergents, dyestuff intermediates, and furan derivatives. Propanesultone is also a potent carcinogen. The carcinogenic action of a mutagen is related to its ability to break chromosomes. Studies on chromosome breakage and carcinogenicity in relation to chemical reactions _in vitro_ have indicated that biological effects depend qualitatively and quantitatively on the alkylating ability of the compound.

12. Hydroxylamine $NH_2OH$

Hydroxylamine is widely used in the transformation of organic compounds to derivatives which in turn may be intermediates in pharmaceutical or other industrial syntheses of complex molecules. One of the largest commercial applications of hydroxylamine (estimated at over 100 million pounds) is in the synthesis of coprolactam, the raw material for nylon 6. Hydroxylamine is used in the preparation of paper pulp brightening and discoloration, in photographic developers and emulsions, as stabilizer of adhesives from glue and lignosulfates, and in the preparation of oximes which are important as antishining agents in paints and for their complexing action with metallic ions. Hydroxylamine is mutagenic in transforming DNA, bacteriophage (S13, ΦX174, and T4), _Neurospora_, and _S. pombe_, and induces chromosome aberrations in human chromosomes, cultured Chinese hamster cells, mouse embryo cells, and _Vicia faba_. N- and O-methyl hydroxylamines are mutagenic in transforming DNA of _B. subtillis_ and _Neurospora_ and induce chromosome aberrations in Chinese hamster cells. It should be noted that hydroxylamine and certain derivatives of hydroxylamine (e.g., N- and O-methyl hydroxylamines), as well as hydrazine, have in common the ability to interact specifically with pyrimidines under certain conditions, including pH, oxygen tension, and concentration of the reagent (their mutagenicity is markedly dependent upon the above conditions).

13. Hydrazine and Derivatives

Hydrazine $H_2NNH_2$

Unsym-1,1-dimethyl-hydrazine $(H_3C)_2NNH_2$

Sym-1,2-dimethyl-hydrazine $CH_3-NH-NH-CH_3$

Hydrazine and its derivatives possess a broad spectrum of utility in photography, preservatives, metal processing, the preparation of agricultural chemicals (e.g., maleic hydrazide, 3-amino triazole and -hydroxyethylhydrazine), medicinals (e.g., isonicotinyl acid hydrazide, nitrofurazone, phenylhydrazine), textile agents, explosives, fuels, and plastics. The yearly consumption of hydrazine in the U. S. in 1964 was estimated at 35 million pounds and is projected to the level of approximately 45 million pounds in 1970. Unsymmetrical 1,1-dimethylhydrazine (UDMH) is prepared by the electrolytic reaction of dimethylnitrosamine and, in addition to its primary use in rocket fuels, it has patent applications as a solvent for acetylene and as a gasoline additive. The induction of chromosome breaks by hydrazine and its derivatives and of mitotic recombination in *Saccharomyces cerevisiae* by sym-dimethylhydrazine has been described.

14. Diazomethane $CH_2=\overset{\oplus}{N}=\overset{\ominus}{N}$

Diazomethane is a very powerful methylating agent for acidic compounds such as carboxylic acids, phenols, and enols, and hence is both an important laboratory reagent and has industrial utility. It is usually prepared in ethereal solution by the decomposition with alkali of a variety of mutagenic N-nitroso compounds, e.g., respective derivatives of urea, urethan, nitroguanidine, and p-toluenesulfonamide. Diazomethane is carcinogenic in rats and mice and is reported to be responsible for the carcinogenicity of many nitroso compounds and related compounds. It is mutagenic in *Drosophila*, *Neurospora*, and *Saccharomyces cerevisiae*, and established to be the principal agent in the mutagenesis of nitroso compounds, e.g., nitrosoguanidine in *E. coli*.

15. Ethylenimine

$$\underset{\text{NH}}{\overset{CH_2\text{—}CH_2}{\diagdown\ \diagup}}$$

Ethylenimine, available in commercial quantities, is an extremely reactive compound that (a) reacts with many organic functional groups containing an active hydrogen to yield an aminoethyl derivative or products derived from them and (b) undergoes ring-opening reactions similar to those undergone by ethylene oxide. Thus ethylenimine, because of its high degree of reactivity, exhibits actual or potential utility in a broad and expanding range of applications including: textiles (crease-, flame-, water-, and shrink-proofing), agricultural chemicals (e.g., insecticides, chemosterilants (preparation of tepa, metepa, and apholate), and soil conditioners), chemotherapeutics (triethylene melamine), petroleum products, and synthetic fuels. Ethylenimine is carcinogenic and has been shown to cause mutations in *Drosophila*, *Neurospora*, wheat, barley, and *Saccharomyces cerevisiae*, and chromosome aberrations in cultured human cells, mouse embryonic skin cultures, Crocker mouse Sarcoma 188, and root tips of *Allium cepa*.

16. Dimethylnitrosamine $(H_3C)_2\text{—N—NO}$

DMN

Dimethylnitrosamine is representative of a category of compounds of great activity and considerable diversity of action, especially as carcinogens. Their occurrence, whether as synthetic derivatives, natural products, or accidental products in food processing or tobacco smoke, is of significant environmental concern. A number of nitrosamines have been patented for use as gasoline and lubricant additives, antioxidants, and pesticides. DMN is used primarily in the electrolytic production of the hypergolic rocket fuel 1,1-dimethylhydrazine. Other areas of utility include the control of nematodes, the inhibition of nitrification in soil, plasticizer for acrylonitrile polymers, and its use in active metal anode-electrolyte systems for high-energy batteries. The nitrosamines have been shown to induce a great variety of tumors at different sites in many species. It seems established that the dialkylnitrosamines are metabolized *in vivo* to yield an active alkylating agent which can be diazoalkane itself, an active alkene, or an alkyl carbonium ion derived from it by further decomposition. It is possible that nitrosamines occur in foods since many foods contain large amounts of amines and small quantities of nitrite as preservative. The presence of nitrosamines has been suggested in tobacco, tobacco smoke, fish meals, wheat kernels, flour, diary products (milk and cheese), in various smoked fish, meat, and mushrooms, and in the fruit of a solonaceous bush (*Solanum incanum*), the juice of which is used to curdle milk (this provides the chief source of sustenance of the Bantus in localized areas of the Transhei where there is a high incidence of esophageal cancer). With regard to the mutagenicity of nitroso compounds, it is of note that the nitrosamines which are believed to require enzymic decomposition before becoming active carcinogens (e.g., dimethyl- and diethylnitrosamines) are mutagenic in *Drosophila* and *Arabilopsis thaliana* and inactive in microorganisms such as *E. coli*, *Serratia marscesens*, *S. cerevisiae*, and *Neurospora* (but active in *Neurospora* in the hydroxylating model of Udenfriend or in the presence of oxygen).

17. Urethan

Ethyl carbamate

$$H_2N-\underset{\underset{O}{\|}}{C}-OC_2H_5$$

Urethan and its derivatives are widely used in the plastics industry as monomers, co-monomers, plasticizers, and fiber and molding resins, in textile finishing, in agriculture as herbicides, insecticides, insect repellants, fungicides, and molluscicides, and in the pharmaceutical industry as psychotropic drugs, hypnotics, sedatives, anticonvulsants, miotics, anesthetics, and antiseptics. Urethan is a pulmonary carcinogen in mice and rats, induces carcinoma of the forestomach in hamsters, and is teratogenic in mice, hamsters, fish, and amphibia. It is mutagenic in *Drosophila*, bacteria, and plants, and induces chromosome aberrations in *Oenothera*. Alkyl congeners of urethan, e.g., methyl-, propyl-, and butyl-carbamate are active mutagens in bacteria.

N-hydroxy-urethan, a mammalian metabolite of urethan, induces chromosome aberrations in *Vicia faba* and mammalian cells in culture and inactivates transforming DNA. (Chromosome breaks and large chromosome alterations, but not point mutations, are induced.)

18. Trimethylphosphate

TMP

$$(CH_3O)_2P(=O)-OCH_3$$

Trimethylphosphate is a methylating agent largely employed as a low-cost gasoline additive (at a concentration of approximately 0.25 g per gallon) and in the preparation of organophosphorus insecticides, polymethyl polyphosphates, as a flame retardant solvent for paints and polymers, as a catalyst in the preparation of polymers and resins, and has recently been proposed as a food additive for stabilizing egg whites. TMP is a chemosterilant in rodents (probably related to *in vivo* alkylation), induces reverse mutations in *Neurospora*, and induces dominant lethal mutations in mice. Information regarding the concentration of unreacted TMP (as well as possible biologically active pyrolysis products) in automobile exhaust would contribute to a better evaluation of the potential human hazard involved in its use.

19. Hydrogen peroxide $H_2O_2$

The production of hydrogen peroxide in 1968 was estimated at approximately 75,000 tons. The major areas of utility of hydrogen peroxide include the bleaching of cotton textiles and wood and chemical (kraft and sulfite) pulps; the oxidation of a variety of important organic compounds, such as soybean and linseed oils and related unsaturated esters, to the epoxides for use as plasticizers and stabilizers for polyvinyl chloride; the preparation of organic peroxides (e.g., peroxy acids, hydroperoxides and diacylperoxides); blowing agent for the preparation of foam rubber, plastics, and elastomers; bleaching, conditioning, or sterilization of starch, flour, tobacco, paper, and fabric; and germicidal and cosmetic applications. Hydrogen peroxide is mutagenic in *Staphylococcus aureus*, *E. coli*, and

*Neurospora*, induces chromosomal aberrations in strains of ascites tumors in mice and in *Vicia faba*, and inactivates transforming DNA.

## 20. Organic Peroxides

Tert-butylhydroperoxide $(CH_3)_3COOH$

Cumene hydroperoxide $C_6H_5$—$C-(CH_3)_2OOH$

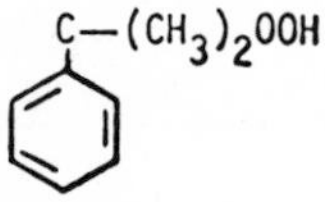

Succinic acid peroxide $HOOC-CH_2CH_2-C(=O)-O-OH$

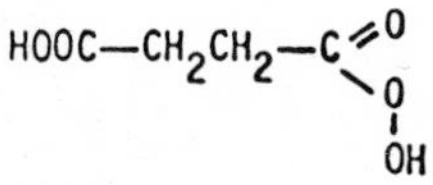

Disuccinyl peroxide $COCH_2CH_2COOH$—O—O—$COCH_2CH_2COOH$

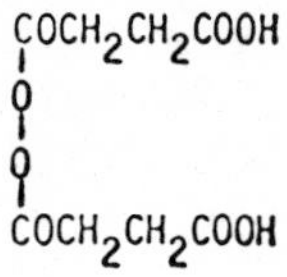

Di-tert-butylperoxide $(CH_3)_3C-O-O-C(CH_3)_3$

These compounds are the major organic peroxides which are extensively (and principally) used in the polymer industry as initiators for the free-radical polymerizations and/or copolymerizations of vinyl and diene monomers, curing agents for elastomers and resins, and cross-linking agents for polyolefins. They have been used in the bleaching of various materials such as flour, gums, waxes, fats, and oils. Together with approximately a dozen other organic peroxides, their consumption in the United States in 1968 amounted to approximately 10 million pounds. The organo peroxides above are used in commercial formulations in concentrations of 70%-99% liquid or powder and also in multiple combinations of mixtures of peroxides. Tert-butylhydroperoxide is mutagenic in *Drosophila*, *E. coli*, and *Neurospora*, and induces chromosome aberrations in *Vicia faba* and *Oenothera*; cumene hydroperoxide is mutagenic in *E. coli* and

*Neurospora*; di-tert-butylperoxide is mutagenic in *Neurospora*. Dihydroxydimethyl peroxide, generally prepared via the reaction of formaldehyde and hydrogen peroxide, has been postulated as a component in polluted air, is mutagenic in *Drosophila*, and exhibits radiomimetic effects on roots of *Vicia faba*.

## SECTION III. FOOD AND FEED ADDITIVE AND NATURALLY OCCURRING MUTAGENS (and Related Degradation Products)

### 1. Caffeine

1,3,7-Trimethylxanthine
Methyltheobromine

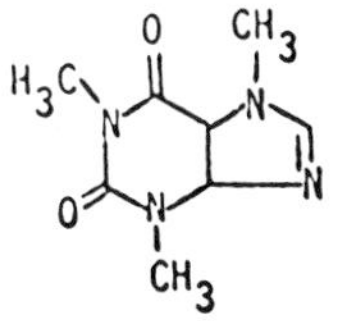

Caffeine is found in the extensively used beverages coffee, tea, cocoa, and mate, as well as in some "soft drinks," particularly the cola-flavored drinks containing about 2% caffeine made from nuts of the tree *Cola acuminata*. In addition to its large consumption in beverages, caffeine has broad medical utility, e.g., with antihistamines to combat motion sickness, with analgesics or ergot alkaloids for relief of tension or migrain headaches, as a stimulant, and as a coronary artery dilator. Evidence is conflicting regarding its mutagenicity. For example, caffeine is weakly mutagenic in *Drosophila*, inducing sex-linked recessive lethals and certain chromosome aberrations; and mutagenic in bacteria (*E. coli*), inducing back mutations to phage resistance and streptomycin non-dependence, fungi (*Ophiostoma multiannulatum*), human tissue culture cells, plants (onion-tip roots), mice, and bacteriophage T5. In bacteria, mutagenicity is claimed to depend upon DNA synthesis or cell division; in non-dividing bacteria, caffeine is reported to be antimutagenic. Induction of recessive lethal mutations has been reported by some investigators and refuted by others. No significant increase in dominant lethals among the progeny of caffeine-fed male mice has also been observed. The teratogenic effect in mice of caffeine in high doses, as well as the ability of caffeine to penetrate the preimplantation blastocyst in the rabbit, has also been reported.

2. Cyclamate

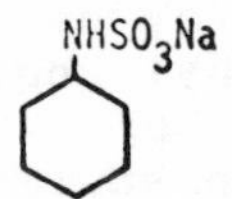


Cyclohexylamine

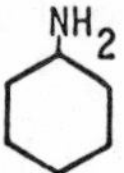


Prior to the recent banning in the United States, Canada, and a number of West European countries of cyclamates (due to induction of bladder tumors in the rat in chronic feeding studies), the projected consumption of cyclamates in 1970 was 21 million pounds. It is of interest to note, however, that in 1963 only 4% of the soft drinks were marketed with cyclamate, compared to 15% in 1968. In addition, until recently, calcium and sodium cyclamate were extensively used alone or in combination with saccharin as non-caloric sweeteners in a wide variety of food products such as beverages, cereals, and bakery products. Cyclamate induces chromosome breakage in onion-root tips, in human leukocytes _in vitro_, and in both leukocyte and monolayer cultures from human skin and carcinoma of the larynx. The synergistic radiomimetic effects of caffeine, alcohol, and sucaryl (which contains 8% sodium cyclamate and 0.8% saccharin) in onion-root tips has also been reported. It is estimated that approximately 30% of all people who consume cyclamate convert it to cyclohexylamine. Other metabolites present in human urine following cyclamate consumption include N-hydrocyclohexylamine, cyclohexanone, and cyclohexanol. Cyclohexylamine induces chromosome breaks _in vitro_ in rodent cells in culture as well as _in vivo_ in rat spermatogonia.

3. EDTA

Ethylenediamine tetraacetic acid

$$(HOOCCH_2)_2NCH_2CH_2N(CH_2COOH)_2$$

Ethylenediamine tetraacetic acid and its alkali salts (e.g. disodium and calcium disodium salts) are widely used as sequestrants in food systems. They are of value in antioxidant systems due to their property of forming poorly dissociable chelate complexes with trace quantitities (0.1 to 5 ppm) of divalent and trivalent metals such as copper and iron in fats and oils. By chelating these metals, pro-oxidant catalytic effects are eliminated or minimized, and in most cases the combined effect of these sequestrants and antioxidants such as the hindered phenols BHT and BHA is synergistic, resulting in enhanced shelf life when used in foods containing fats and oils, e.g., mayonnaise, oleomargarine, and salad dressing. Other areas of utility of EDTA and its derivatives include the purification of antibiotics and in pesticidal compositions, plant nutrients, plant growth regulation, water treatment, and bactericidal and germicidal compositions. EDTA has been shown to induce chromosome aberrations and breakage in _Tradescantia paludosa_, production of mitotic abnormalities in onion roots, chromosomal changes in _Drosophila_, _Vicia faba_, and _Hordeum_, and possesses a synergistic effect in the production of chromosome aberrations in _Vicia faba_.

4. Allylisothiocyanate

$CH_2{=}CHCH_2NCS$

Volatile oil of mustard

The natural source of allylisothiocyanate is sinigrin, one of the major glucosinolates in a variety of plants such as cabbages, kale, brussel spouts, cauliflower, broccoli, kohlrabi, and in horseraddish and white, brown, and black mustard. Allylisothiocyanate is released from sinigrin by enzymic hydrolysis and is used as a food additive (body, modifier) of hot sauces, relish flavors, salad dressings and synthetic mustard. It also has utility as a meat preservative and plant growth regulator. Allylisothiocyanate is mutagenic in _Drosophila_ and _Ophiostoma_ and induces chromosome aberrations in _Drosophila_ and _Allium cepa_.

5. Sodium Nitrite

$NaNO_2$, $HNO_2$

Nitrous Acid

Sodium nitrite is extensively used as a preservative in meat, fish, and cheese. The effectiveness of nitrite, a weak acid salt, is related to pH. The possible formation of nitrosamines from amines present in or derived from the diet occurs by reaction with nitrous acid at pH 4. In man, gastric juice attains a pH of 1.1. Such high concentrations of hydrogen ions give rise to the nitrosyl cation $NO^+$ which is a highly reactive nitrosating agent. The presence of meat, myoglobin, or hemoglobin serve to neutralize this cation by reaction with $Fe^{++}$. However, fish and cheese do not provide this neutralization, hence permitting the nitrosation of amines to possible carcinogenic and mutagenic nitrosamines. Nitrous acid is mutagenic toward transforming DNA, _E. coli_, _Salmonella typhimurium_, _Neurospora_, _Saccharomyces cerevisiae_, _S. pombe_, phage (S13, ΦX174, T4, and T2), _Aspergillus nidulans_, _A. niger_, _A. amstelodami_, and tobacco mosaic virus. Evidence to date indicates that deamination of adenine and cytosine is primarily responsible for the mutagenicity of nitrous acid. In addition to deamination, nitrous acid produces interstrand cross-links within the DNA molecule which interfere with the replication of DNA, hence producing the liklihood that the induction by nitrous acid of deletion mutations is a consequence of cross-linking.

6. Sodium Bisulfite $NaHSO_3$

Sodium bisulfite is used as a bacterial inhibitor in wine, brewing beverages, fruit juices; and as a preservative in dried fruits. The bisulfite anion reacts rather specifically with uracil and cytosine, most likely within single-stranded regions in the case of polynucleotides. It is reported to be mutagenic in _E. coli_ and the bacteriophages lambda and T4.

7. Cycasin

Methylazomethane-β-D-glucoside $CH_3-\underset{\downarrow \atop O}{N}=N-CH_2O$-β-D-Glucopyranosyl

Methylazoxymethanol $CH_3-\underset{\downarrow \atop O}{N}=NCH_2OH$

MAM

Cycasin and its aglycone, methylazoxymethanol (MAM) are extractable from nuts, seeds, and roots of cycad plants, primarily from the very widespread species Cycas circinalis and Cycas revoluta. Cycads are often utilized as food and medicines, principally in the tropics and subtropics. (The cycad nuts are a source of starch.) Cycasin has been shown to be hepatoxic and carcinogenic in rats. It is known that cycasin must be deglucosylated before it is carcinogenic and that intestinal flora provides the β-glucosidase activity necessary for this hydrolysis. The aglycone of cycasin (MAM) is the proximate carcinogen since it causes neoplasms after subcutaneous and intraperitoneal injection, as well as feeding, while cycasin causes tumors in mature animals only if fed. In addition to its hepatoxic and carcinogenic effects, MAM is teratogenic to the golden hamster. MAM is a potent mutagen in Drosophila melanogaster and Salmonella typhimurium. Both cycasin and MAM have been found to increase the mutant frequency of Salmonella typhimurium histidine auxotrophs when tested in the host-mediated assay. MAM also has been found to induce chromosome aberrations in Allium seedlings equivalent to 200 r of x-rays.

8. Pyrrolizidine Alkaloids

Heliotrine

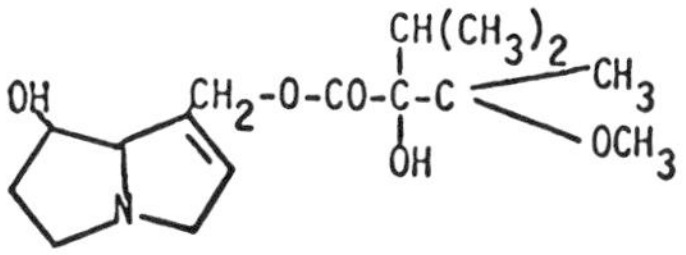

Lasiocarpine

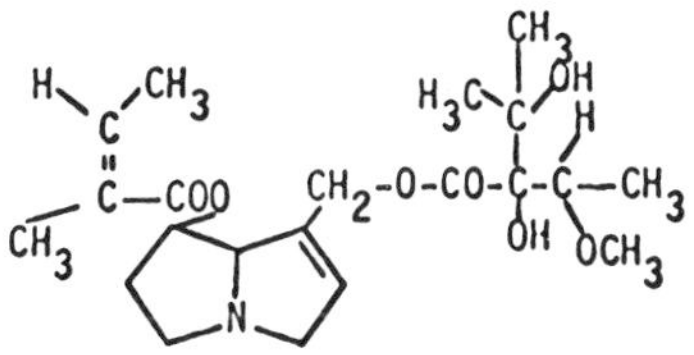

Monocrotaline

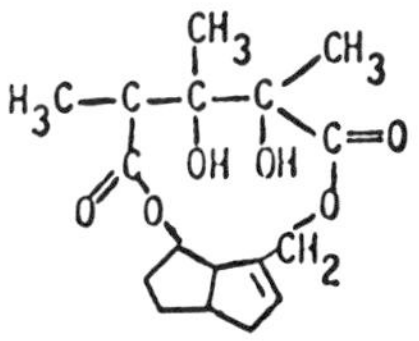

Alkaloids of the pyrrolizidine class are found in members of the Senecio, Crotolaria, Amsinkia, and other genera which are widely distributed throughout rangelands and are hepatoxic to livestock. Many of the plants containing such alkaloids have been and are being traditionally used as herbal folk medicines for various disorders. Pyrrolizidine alkaloids were among the first to be suggested as possible natural etiological factors in the high incidence of liver diseases (kwashiorkor, cirrhosis, and primary liver tumors) in the tropics and subtropics. It has been suggested that the pyrrolizidine alkaloids are not hepatoxic per se, but are transformed in the liver by ring dehydrogenation to pyrrole-like derivatives which react with tissue constituents to form soluble "bound pyrroles" (subsequently excreted in the urine) and insoluble "bound pyrroles" taken up in tissues where reaction with sulfhydryls (or with DNA in a manner as to inhibit synthesis of messenger RNA) would occur more readily than with the intact alkaloid precursor. The mutagenicity of the pyrrolizidine alkaloids has been demonstrated in *Drosophila melanogaster* and *Aspergillus nidulans*, as well as their induction of chromosome aberrations in *Drosophila* and *Allium cepa*.

9. Bracken Fern

Bracken fern, *Pteridium aquilinium*, has been often associated with the fatal poisoning of cattle in various parts of the world. (This poisoning has been produced by feeding fresh foods, sun-dried bracken, or powdered rhizomes mixed with an otherwise adequate diet and an alcoholic fraction of the bracken plant.) The bracken plant is known to contain radiomimetic activity, and produces adenocarcinoma of the intestinal mucosa when fed to rats. It is of interest to note that, in Japan, a plant similar to bracken fern, zen mai (*Osmunda Japonica*), is used as a vegetable and seasoning, and it has been suggested that bracken may be confused with zen mai and hence might contribute to the relatively high incidence of stomach cancer found in Japan. It has been suggested that the observed carcinogenicity and alkylating potential of bracken fern when fed to cattle may be due to dimethylsulfonium propionic acid dihydrochloride. Bracken fern is mutagenic in *Drosophila* and mice and induces sterility in quail (due to direct action on spermatogenesis and/or the expression of dominant lethal mutations.

10. Aflatoxin $B_1$

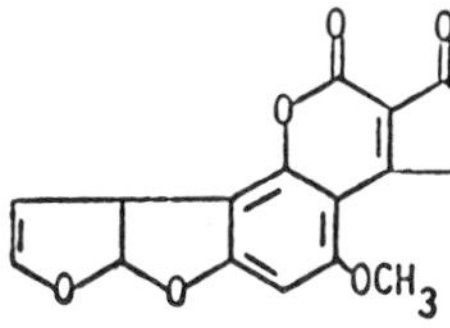

Aflatoxins (consisting of 8 compounds of related molecular configurations) are toxic mold metabolites produced by a limited number of strains of a few fungi, e.g., *Aspergillus flavus*, *A. parasiticus*, and *Penicillium puberulum*, and whose occurrence in feeds and foods (e.g., groundnut, peanuts) and high order of carcinogenic potency in many species (principally aflatoxin $B_1$) has been well documented. There is considerable speculation as to whether the aflatoxins may be involved in the etiology of human liver disease, including primary carcinoma. Aflatoxin is mutagenic in the dominant lethal test in mice and *Vicia faba* and suppresses mitosis in cultured human diploid and heteroploid embryonic lung cells, and has exhibited a facility for binding to DNA, affecting DNA polymerase of *E. coli*, and inhibiting DNA synthesis and giant cell formations in tissue culture. It has been suggested that aflatoxin $B_1$ combines with DNA and forms complexes, indicative of base-pair intercalation. Aflatoxin $B_1$ is also a powerful inhibitor of RNA synthesis.

11. Polynuclear Aromatic Hydrocarbons

3,4-benzpyrene

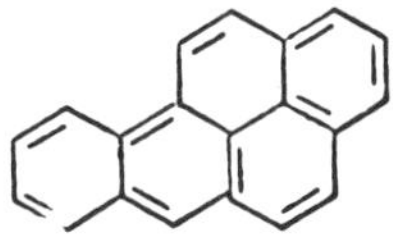

1,2,5,6-dibenz-anthracene

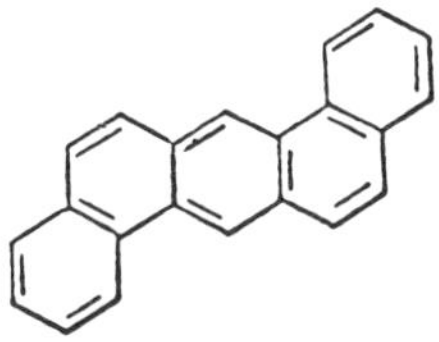

3,4-benzpyrene, the most ubiquitous and potent known carcinogenic agent in polluted air, is primarily produced as a result of the incomplete combustion of solid, liquid, and gaseous fuels.

However, it is also found in a broad range of environmental sources including cigar, cigarette, and tobacco smoke condensate, rubber tire dust, refined oils (mineral oil), freshly mined asbestos, marine fauna and flora, commercial wax, smoked foods, and in foods such as roasted coffee, baked bread and biscuits, margarine and mayonnaise, and oranges. 3,4-benzpyrene is mutagenic in Drosophila, E. coli, and mice (dominant lethal test). 1,2,5,6-dibenzanthracene has been isolated from cigarette smoke, mineral oils, smoked meat and fish, and oranges. It is mutagenic in Drosophila, Neurospora, E. coli, and mice, and it (as well as 3,4-benzpyrene) induces mitotic gene conversion in Saccharomyces cerevisiae and has also been found to be carcinogenic.

12. Patulin

4-hydroxy-4H-furo-
(3,2-C)pyran-2(6H)one

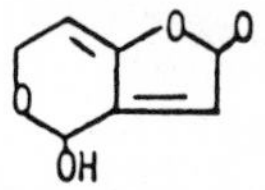

Patulin is an α,β-unsaturated lactone antibiotic derived from the metabolism of several species of Aspergillus and Penicillium (e.g., A. clavatus, A. clavifome, P. patulum, P. expansum, P. meliniis, and P. lencopus.) Some of these fungal species are likely contaminants of foods. Patulin has been isolated from flour, field crops, apples, and fruit juices. Patulin, which is carcinogenic, has been shown to induce petite mutations in Saccharomyces cerevisiae and chromosome aberrations in avian eggs during mitosis and in human leukocyte cell culture.

## SECTION IV. DRUG MUTAGENS

Among the drugs that have been found to induce point mutations and chromosome aberrations are agents illustrative of a variety of classes of chemical structure, e.g., alkylating agents, antibiotics, acridines, folic acid antagonists, etc., as well as possessing a broad spectrum of utility, e.g., antineoplastic, antibacterial, antimalarial, antipyretics, sedatives, etc.

## 1. Alkylating Agents

By far the largest category of utility of the drugs that have been found to be mutagenic is in the area of cancer chemotherapy, with the most potent agents being alkylating agents, a number of which are discussed below.

### a. Cyclophosphamide

Endoxan
N,N-bis(2-chloroethyl)-N,O-propylene phosphoric acid ester diamide

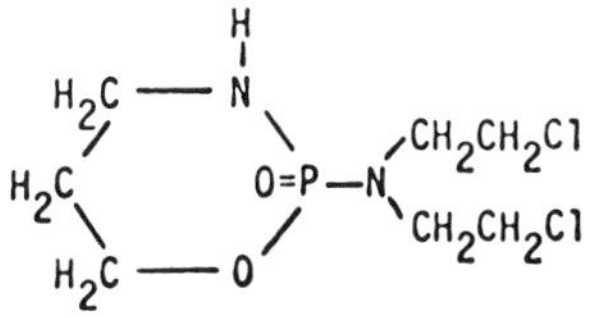

Cyclophosphamide exhibits a significantly greater selectivity against many kinds of tumor cells than other cytostatic agents of the nitrogen mustard series. It is an antitumor agent with an inert transport moiety and is activated only *in vivo* to nor-$HN_2$. It has been shown to be teratogenic in the rat and mutagenic in mice and *Drosophila*, as well as to induce chromosome aberrations in Chinese hamster bone marrow, murine, and human cells *in vivo*.

### b. Trenimon

2,3,5-tris-ethylenimino-1,4-benzoquinone

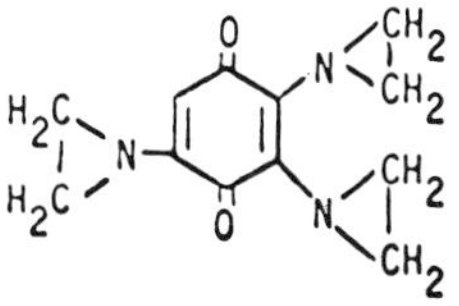

Trenimon is mutagenic in mice, *Drosophila*, and human leukocyte chromosomes *in vitro*, as well as causing chromosome aberrations in barley and *Vicia faba*.

### c. Myleran

Busulfan
Methanesulfonic acid tetramethylene ester

$CH_3SO_2O(CH_2)_4OSO_2CH_3$

Myleran is mutagenic in mice, *Drosophila*, and barley, induces chromosome aberrations in human leukocytes, *Vicia faba*, and most spermatogonia, and has been implicated in human teratogenesis.

d. Nitrogen Mustard

Methyl di(2-chloro-ethyl)-amine — $(ClCH_2CH_2)_2NCH_3$

Nitrogen mustard is the first of the nitrogen mustards to be introduced into clinical medicine and has been more avidly studied than any of its congeners. The nitrogen mustards are chemically very reactive, electrophilic compounds that react with and alkylate nucleophilic substances and such biologically important moieties as phosphate, amino, sulfhydryl, hydroxyl, imidazole, and carbonyl groups. The mutagenicity of nitrogen mustard in mice, *Drosophila*, and a variety of microorganisms has been demonstrated, as well as its induction of chromosome aberrations in *Vicia* and *Allium cepa*.

e. Other Alkylating Agents

Tretamine
TEM
Triethylene melamine

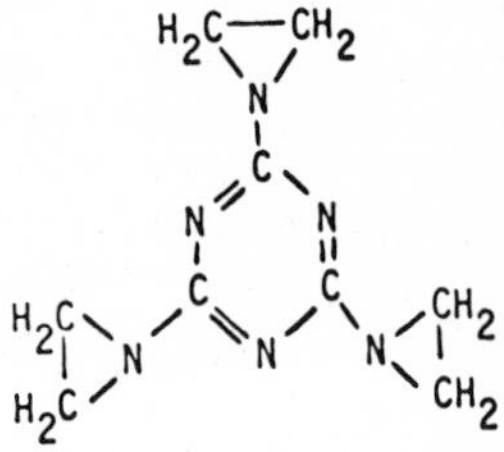

NOR-$HN_2$
2,2'-dichloroethyl-amine — $(ClCH_2CH_2)_2NH$

$HN_3$
Tris(2-chloroethyl)-amine — $(ClCH_2CH_2)_3N$

Nitromin
N-oxide-mustard — $(ClCH_2CH_2)_2N(\rightarrow O)—CH_3$

Sulfur mustard
Bis(2-chloroethyl)-sulfide — $S(CH_2CH_2Cl)_2$

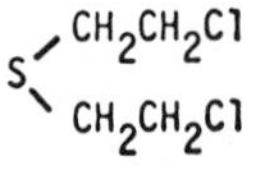

These alkylating agents have been used in cancer chemotherapy.

2. Antibiotics

The streptomyces-derived antibiotics such as mitomycin C, azaserine, streptonigrin, and phleomycin are DNA inhibitors and have been shown to be of utility as antineoplastic agents.

a. Mitomycin C

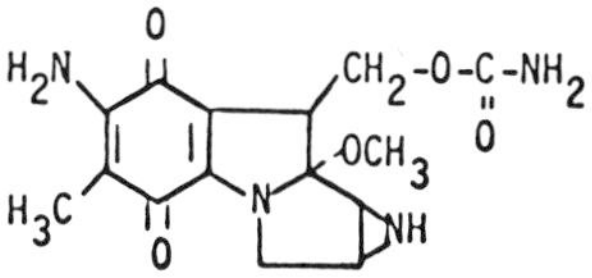

Mitomycin C is a bifunctional alkylating agent and acts to cross-link the two backbones of the double helix of DNA. (It is considered as a derivative of urethan and of ethylenimine and is biologically inactive in its natural state but becomes a mono- and bifunctional alkylating agent upon chemical or enzymic reduction.) It is mutagenic in _Drosophila_ and Habrobracon sperm and induces chromosome aberrations in cultured human leukocytes and _Vicia faba_.

b. Streptonigrin

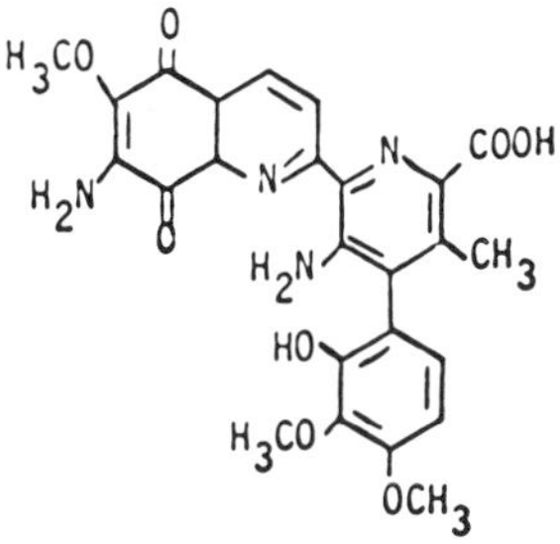

Streptonigrin is an extremely potent and selective inhibitor of bacterial DNA synthesis and has been found to be mutagenic in the ascomycete _Ophiostoma multiannulatum_ and to induce chromosome aberrations in cultured human leukocytes and in _Vicia faba_.

c. Azaserine o-diazoacetyl-L-serine

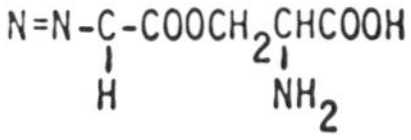

N=N-C(H)-COOCH$_2$CH(NH$_2$)COOH

Azaserine is a glutamine antagonist which inhibits purine synthesis in some bacteria and is mutagenic in *E. coli* and T2 phage and induces chromosomal aberrations in root tips of *Allium cepa*, *Vicia faba*, and *Tradescantia patudosa*.

3. Folic Acid Antagonists

a. Aminopterin
4-aminofolic acid

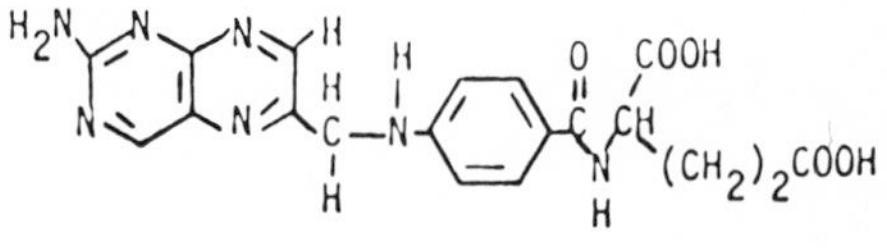

Methotrexate
Amethopterin
4-amino-10-methyl-folic acid

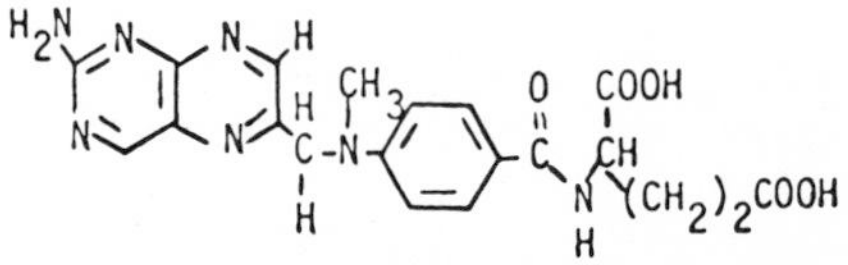

Folic acid antagonists such as aminopterin and methotrexate have also been widely used in cancer chemotherapy. Methotrexate is an antifolic compound that acts by competitive inhibition of folate reductase thus acting directly on thymidine biosynthesis. It has been extensively used in the treatment of acute leukemia and choriocarcinoma, as well as in the therapy of severe psoriasis, and is an immunosuppressive agent in organ transplants. Both aminopterin and methotrexate cause chromosomal breakage in cultured human lymphocytes and are teratogenic in birds and mammals.

b. Daraprim

2,4-diamino-5-(p-chlorophenyl)-6-ethylpyrimidine

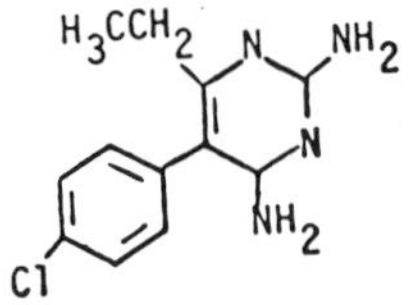

Daraprim is a folic acid antagonist used as an antimalarial and has been found to induce chromosomal abnormalities in man.

4. Vincristine

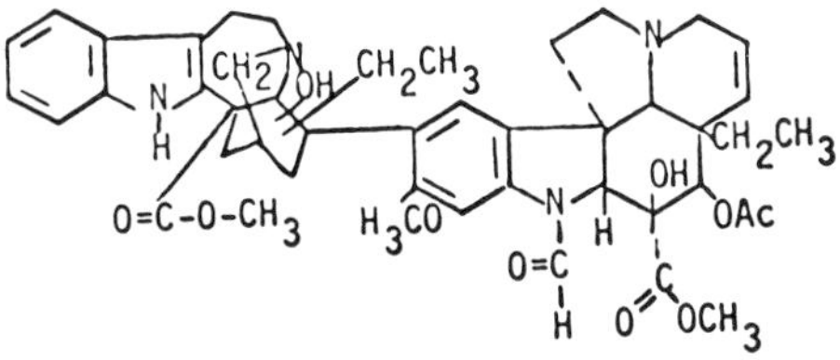

Vinblastine
Vinca leukoblastine

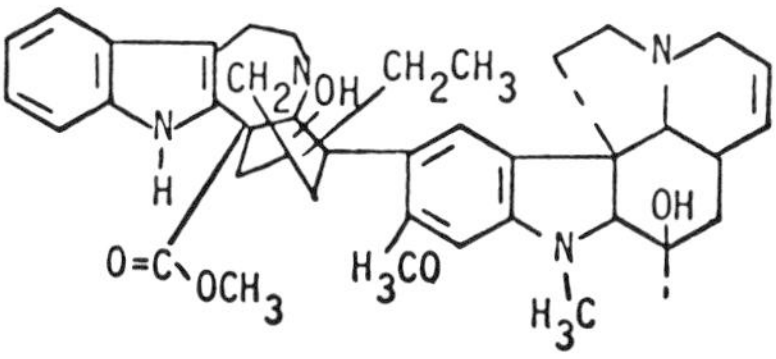

The vinca alkaloids such as vincristine and vinblastine derived from the periwinkle plants Vinca rosa, Linn have received extensive clinical investigation as cytostatic agents primarily for use in Hodgekin's disease and choriocarcinoma. Both vincristine and vinblastine induce chromosome aberrations in plants and mammalian cells in culture.

5. Acridine Derivatives

Acriflavine
A mixture of 3,6-diamino-10-methylacridinium chloride and 3,6-diamino-acridine

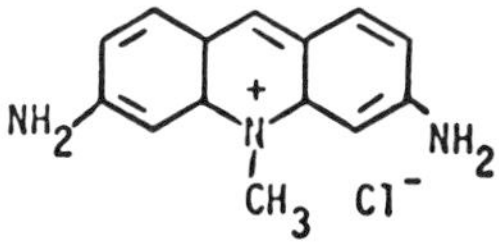

+ Proflavine

Proflavine
3,6-diaminoacridine

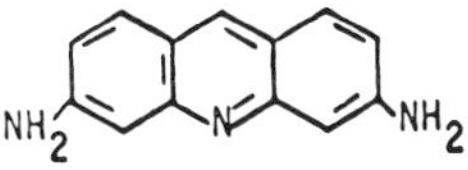

5-aminoacridine

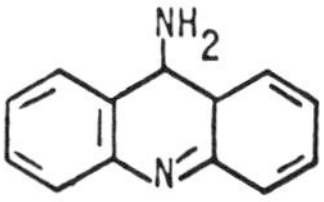

Atabrine
Quinacrine
6-chloro-9(4-diethyl-amino-1-methylbutyl-amino)-2-methoxy acri-dine dihydrochloride

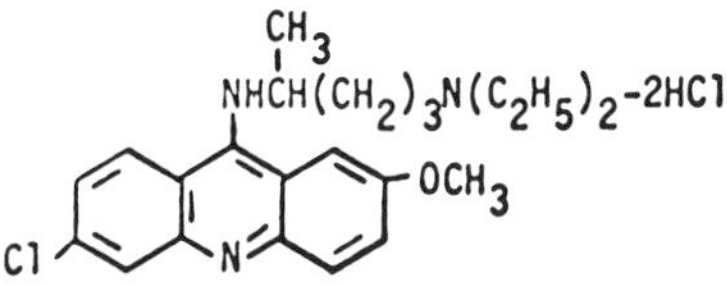

$NHCH(CH_3)(CH_2)_3N(C_2H_5)_2 \cdot 2HCl$

The acridine dyes and derivatives generally available for medical and veterinary use as antiseptic and bactericidal agents are acriflavine, proflavine, and 5-aminoacridine. Proflavine induces mutations in phages T2 and T4 and chromosome breakage in human cells; acriflavine is mutagenic in *Drosophila*, *E. coli*, yeast, phage T4, and *Saccharomyces cerevisiae*. The antimalarial drug atabrine (quinacrine) is also an acridine derivative which is highly mutagenic to *Drosophila*, bacteria, and bacteriophage.

6. Miscellaneous Mutagenic Drugs

a. Hydroxyurea
N-hydroxyurea

$$H_2N\text{-}\underset{\underset{O}{\|}}{C}NHOH$$

Hydroxyurea is an antineoplastic agent with demonstrated utility in acute lymphoblastic and chronic myelogeneous leukemia. It has been shown to inhibit DNA synthesis and induce chromosome aberrations in Chinese hamster cells and mouse embryo cells in culture and in *Vicia faba*, as well as being teratogenic in hamsters, rats, and chick embryo.

b. Urethan
Ethyl carbamate

$$H_2N\text{-}\underset{\underset{O}{\|}}{C}\text{-}OC_2H_5$$

Urethan has been used medically as a mild hypnotic, sedative, and antispasmodic and in the treatment of chronic myeloid leukemia and related blood diseases. The mutagenicity, carcinogenicity, and teratogenicity of urethan has been described above in Section II, Industrial Mutagens.

c. Dithranol
1,8,9-trihydroxy-anthracene

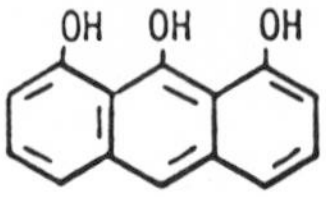

OH OH OH

Dithranol is used in the treatment of psoriasis, ring worm infections, and other chronic dermatoses. It induces petite mutations in yeast *Saccharomyces cerevisiae* and has been shown to have a tumor-promoting property in mice following pre-treatment with 7,12-dimethylbenzanthracene (DMBA). This is of interest because this agent has been used in dermatological practice often after earlier treatment with coal tar.

d. Miracil-D
1-(2-diethylamino-ethylamino)-4-methyl thiazanthone

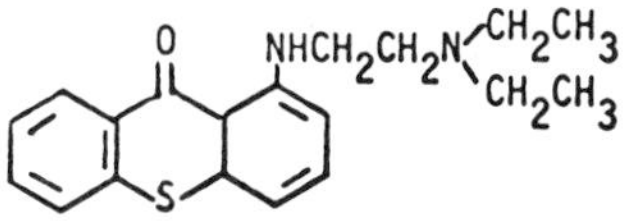

Miracil-D is used in the treatment of schistosomiasis in man. It has been shown to inhibit DNA-directed RNA polymerase and causes complete inhibition of RNA synthesis and partial inhibition of DNA and protein synthesis. Miracil-D is mutagenic in _Drosophila_ and induces acromatic lesions and chromatid breaks in human leukocyte chromosomes _in vitro_.

e. Ethidium bromide
2,7-diamino-9-phenyl-10-ethyl-phenathridium bromide

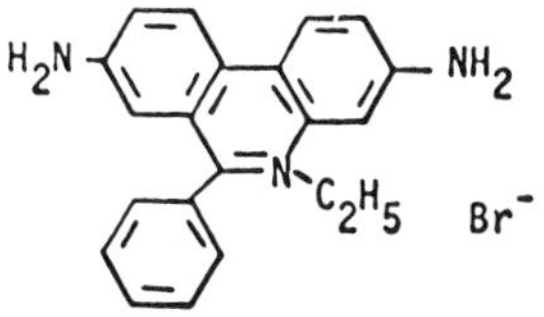

Ethidium bromide is an intercalating dye used as a trypanocide in veterinary medicine. It is probable that the effects of ethidium bromide on nucleic acid syntheses _in vivo_ and _in vitro_ are direct consequences of its physical binding to DNA and RNA. Ethidium bromide induces mutations of yeast mitochondria.

f. Chloral hydrate
Trichloroacetaldehyde monohydrate

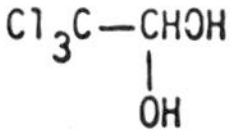

$Cl_3C-CHOH$
$\quad\quad\quad|$
$\quad\quad\quad OH$

Chloral hydrate is used as a hypnotic and sedative. It is also a metabolite of trichloroethylene and is used in herbicidal and industrial synthesis. It has been shown to be mutagenic in bacteria and _Drosophila_.

g. Oxine
8-hydroxyquinoline

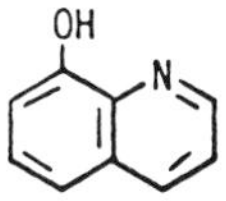

8-hydroxyquinoline is an antibacterial agent and fungistat that is mutagenic in _A. oryzae_ and _A. niger_, induces chromosome aberrations in _Bromus inermis_ root tips, and is carcinogenic in mice.

h. Phenyl butazone
1,1-diphenyl-4-n-butyl-3,5-dioxo-pyrazolidine

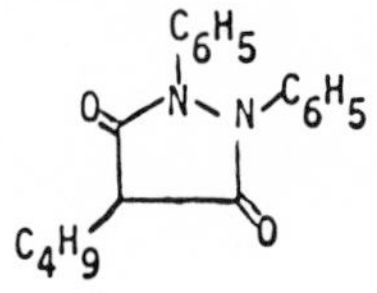

Phenylbutazone is a widely used analgesic, antipyretic, and antiinflammatory agent in human and veterinary medicine that has been found to induce chromosome aberrations in cultured human cells.

i. Meclizine
1-(p-chloro-α phenyl-benzyl)-4-(m-methyl-benzyl)-piperazine

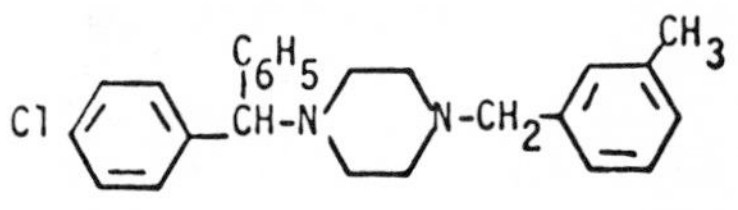

Meclizine is used as an antinauseant and antihistaminic and induces chromosome breakage in the onion root tip.

j. Negram
Nalidixic acid
1-ethyl-1,4-dihydro-7-methyl-4-oxo-1,8-naphthridine-3-carboxylic acid

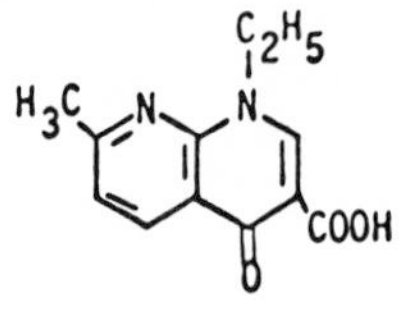

Negram is a bacteriostatic agent used in the treatment of urinary tract infections and has been shown to induce chromosome aberrations of human lymphocytes _in vivo_.

k. Merthiolate
Thimerosal
Sodium methyl mercury thiosalicylate

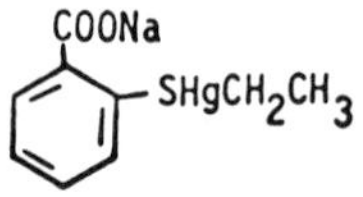

Merthiolate is a widely used topical antiseptic in human and veterinary medicine that has been found to induce chromosome aberrations in _Drosophila_.

l. LSD
Lysergic acid
diethylamide

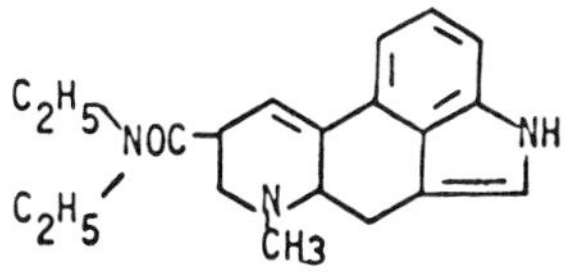

$C_2H_5$ NOC ... NH ... N ... $CH_3$
$C_2H_5$

LSD is one of the most potent abuse hallucinogens known and has been used in the experimental treatment of mental diseases. LSD has been found teratogenic in mice, rats, hamsters, rabbits, and monkeys. The question of the mutagenicity of LSD and its induction of chromosome aberrations _in vivo_ is conflicting. However, the _in vitro_ effect of LSD remains for the most part unquestioned. LSD has been shown to interact directly with DNA, probably by intercalation, causing conformational changes which may serve as a model for the interaction of LSD on the chromosomal material of intact cells.

m. Colchicine

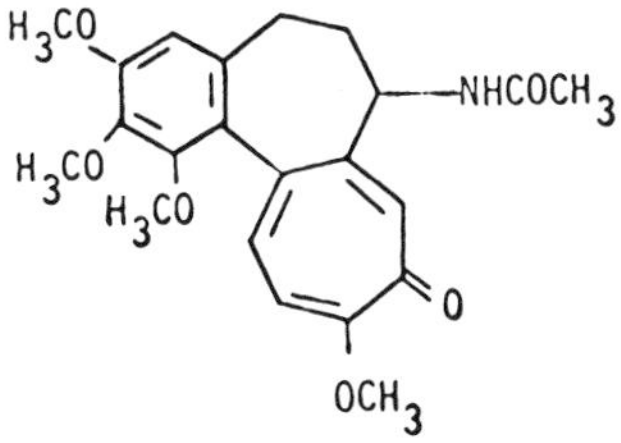

$H_3CO$ $NHCOCH_3$
$H_3CO$ $H_3CO$ O
$OCH_3$

Colchicine, an alkaloid found in the corm and seeds of _Colchicum autamnale_, is widely used in the treatment of acute gout and has also proved effective in the chronic treatment of sarcoid arthritis. It has also been widely employed as an experimental tool in the study of normal and pathological cell growth and the effects thereon of carcinogens, hormones, and other substances, and has been shown to induce chromosomal aberrations in plant and mammalian cells.

n. Hydrogen peroxide $H_2O_2$

Hydrogen peroxide is a widely used topical antiseptic in human and veterinary medicine as well as enjoying extensive use in pharmaceutical preparations, mouthwashes, dentifrices, and sanitary lotions. Its mutagenicity has been described earlier in Section II, Industrial Mutagens.

o. Isoniazid
Isonicotinyl
hydrazide

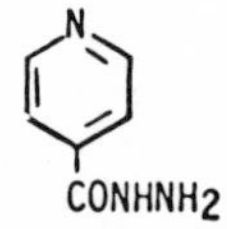


Isoniazid is a widely used antitubercular agent that has been shown to be mutagenic in *E. coli*, as well as tumorigenic and carcinogenic in laboratory animals. Hydrazine, the main metabolite of isoniazid, is mutagenic and has been discussed previously.

p. Chlorpromazine

2-chloro-10-(3-dimethylaminopropyl)-phenothiazine

Chlorpromazine is a widely used sedative and tranquilizer in both human and veterinary medicine. Chlorpromazine and related phenothiazines are employed primarily in two areas, the treatment of psychiatric patients and the treatment of nausea and vomiting. At present there are more than two dozen phenothiazine drugs used in medicine, about half of them for psychiatric conditions. Chlorpromazine has been found mutagenic in the dominant lethal test (Legator, 1971).

## References

Fishbein, L., W. G. Flamm, and H. L. Falk (1970). *Chemical Mutagens*. New York, Academic Press.
Barthelmess, A. (1971). In *Chemical Mutagenesis in Mammals and Man*, F. Vogel and G. Rohrborn, editors. Berlin, Springer-Verlag.
Shaw, M. W. (1971). *Ann. Rev. Med.*, *21*, 409.
Fishbein, L. (1971). *Chromatog. Revs.*, *13*, 83.
Fishbein, L. (1969). *Ann. N. Y. Acad. Sci.*, *163*, 869.
Legator, M. (1971). 2nd Annual Meeting, Environmental Mutagen Society, Washington, D. C., March 21-24.

# MUTAGENICITY OF BIOLOGICALS*

*Warren W. Nichols*

The term biologicals refers to all medicinal preparations of a biologic nature, including vaccines, antitoxins, blood, serums, and antigens. The mutagenic potential of biologicals has received little attention up to the present time, but possible mutagens in biologicals are microorganisms, especially viruses, and a few rather specific chemicals. The microorganisms may be either the biological agents themselves, as in virus vaccines, or contaminants of biologicals such as blood or serums. The chemical agents of concern include preservatives and other additives and also breakdown and metabolic products in the biologicals.

Several viruses, mycoplasma, and some cell products produce chromosome breaks. Chromosome breakage is part of chromosome mutation but, perhaps more important, it is highly correlated with gene mutations. This is manifest in irradiation studies in which amounts of radiation produce similar rates of single hit chromosome aberrations and induced gene mutations (Sparrow, 1961). In chemical studies it is well exemplified by the correlations between gene mutations in a variety of systems and chromosome breaks in mammalian cells in culture (Kihlman, 1966). It is likely that chromosome breaks reflect a spectrum of damage, from the microscopically visible level down to subvisible change, which can include gene deletions. It is unlikely that point mutations (changes in single bases) are a part of this spectrum of damage, since point mutations are specific chemical events that occur by different mechanisms.

---

*Supported by Research Career Award 5-K3-16,749 and General Research Support Grant FR-5582 from the National Institutes of Health; and State of New Jersey Grant-in-Aid Contract M-43.

However, it is possible that point mutations can be produced by errors in repair subsequent to the virus induced defect.

More than twenty viruses have been found to produce chromosome defects. (For recent reviews see Nichols, 1969, 1970; Makino and Aya, 1968; Moorhead, 1970; Stich and Yohn, 1970). Any of these viruses could be present and active as contaminants of blood, serums, or vaccines. The viral contaminant that seems to be most widespread through the use of blood and serums is the hepatitis virus, and chromosome breaks in the blood or bone marrow of patients with clinical hepatitis have been reported on three occasions (El-Alfi et al., 1965; Matsaniotis et al 1966; Mella and Lang, 1967).

Live viruses in vaccines are of more direct concern in causing genetic damage than viruses present as contaminants. Chromosome breaks are associated with three virus diseases for which a virus vaccine is available--measles (rubeola), German measles (rubella), and mumps. Only in the case of measles has the incidence of chromosome breakage in the clinical disease been compared with the incidence in vaccine administration, and the amount of breakage following vaccine administration was far less than that associated with the clinical disease. When the vaccine was administered with gamma globulin, the breakage rate did not exceed the controls (Nichols, 1963). This indicates that fewer chromosome breaks occur with the vaccine than with the natural disease, but the relationship, if any, to gene mutations is less clear.

There might be two explanations for a relationship between chromosome breaks and gene mutations. Figure 1 illustrates a spectrum of damage to chromosomes, from microscopically subvisible lesions to cell death, that might occur during clinical measles. One possible explanation is that after administration of measles vaccine, all events on the graph are reduced, as in Figure 2. In this case, the number of chromosome breaks to be repaired is reduced, and hence the incidence of gene mutations that might be introduced by defective repair mechanisms is also reduced. However, it is possible that, rather than a reduction in all aspects of the

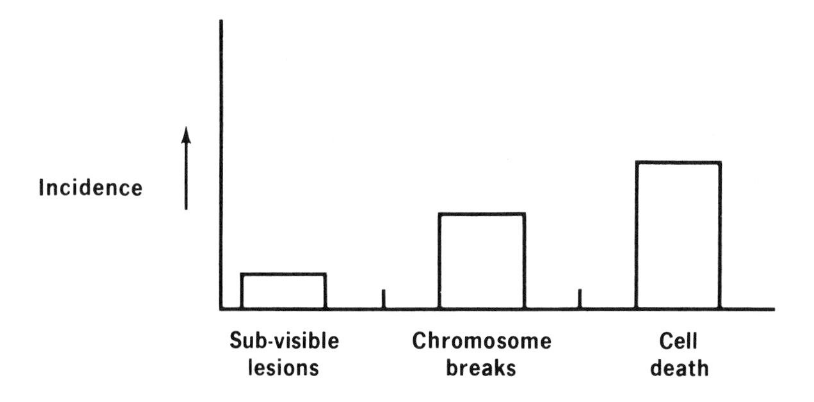


**FIGURE 1. HYPOTHETICAL GENETIC DAMAGE DURING CLINICAL MEASLES**

graph, there is a change in the shape of the graph, as depicted in Figure 3. In this case, gene mutations associated with microscopically subvisible defects might actually increase. At the present time, all of this is speculative. There is no evidence that gene mutations result from virus-induced genetic damage, other than the previously mentioned correlations with radiation and chemical studies. If gene mutations are produced, it will be necessary to determine whether the mechanism of their production is gene deletion or point mutation via aberrant repair mechanisms to evaluate properly these hypothetical possibilities.

Other vaccines that produce chromosome breaks are yellow fever vaccine (Harnden, 1964) and smallpox vaccine (Zur Hausen, 1966). The former involved a study of chromosomes from peripheral leukocytes of patients receiving the vaccine. The latter was in a tissue culture system. Patients successfully vaccinated against smallpox failed to reveal any chromosome abnormalities in their bone marrow metaphase plates (Matsaniotis et al., 1968). Current work in our own laboratory indicates very little if any chromosomal damage with vaccinia virus in tissue culture.

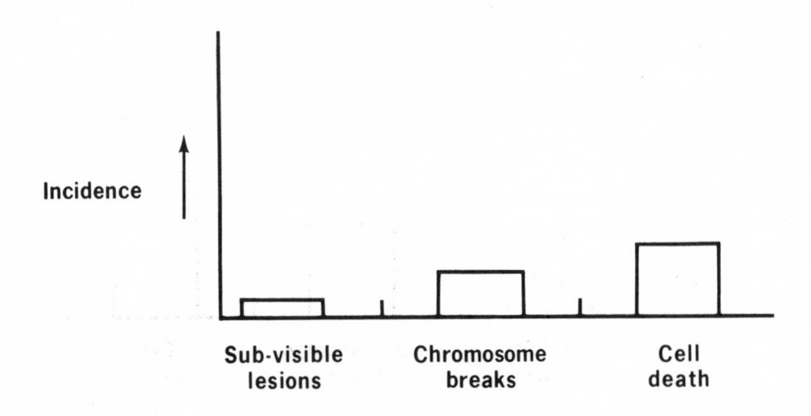


**FIGURE 2. FIRST HYPOTHETICAL GRAPH OF GENETIC DAMAGE OCCURING AFTER ADMINISTRATION OF MEASLES VACCINE**

In addition to chromosome breakage, evidence that viruses are capable of producing genetic changes in cells is found in the virus induced cellular transformations in tissue culture. These systems employ tumor viruses and have been equated with malignancy *in vitro*. A variety of altered cellular characteristics are inherited by subsequent cell generations. In this case, the alteration may be produced by change in the cell's genome or by the acquisition by the cell of a functional viral genome.

The use of viruses as biological insecticides (Ignoffo, 1968) is of concern in relation to genetic hazards, as well as other chronic toxicity, because of the potential for wide human exposure. Viruses are used to control the corn worm and boll weevil by dissemination from airplanes. Biological pesticides offer possible solutions to many important problems, but their potential genetic and other chronic toxicity should be evaluated before use.

Another frequent contaminant of biologicals, the mycoplasma, have produced chromosome breaks in cells in tissue culture (Fogh and Fogh, 1965; Paton et al., 1965; Aula and Nichols, 1967).

Chemicals that can occur in biologicals from the breakdown of cells are exemplified by deoxyadenosine

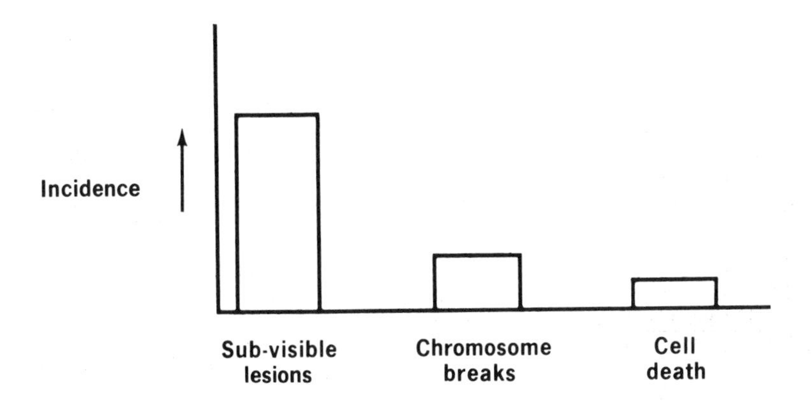


**FIGURE 3. SECOND HYPOTHETICAL GRAPH OF GENETIC DAMAGE OCCURING AFTER ADMINISTRATION OF MEASLES VACCINE**

and cytidine triphosphate. These can result from the breakdown of DNA and RNA, respectively. These chemicals, when added in excess to cells _in vitro_, have resulted in chromosome defects (Kihlman et al., 1963; Nichols et al., 1965). The amounts would not ordinarily be expected to be significant in sera or blood unless there had been considerable cellular breakdown. Genetic risk from sources of this type is probably minimal, but little data exists at this time.

## Conclusion

At the present time the greatest genetic hazards in biologicals are live virus vaccines and microbial contamination, as evidenced by the association of chromosome breaks with a large variety of viruses and mycoplasma. There is currently no firm evidence that chromosome breaks are associated with gene mutations. At present, the clear and often life-saving benefit of many biologicals and the great number of unanswered questions regarding the potential risk would speak for their continued use. However, in order to achieve the greatest possible safety, studies on the degree of genetic risk involved and methods to minimize any risk that exists should progress as rapidly as possible.

## References

Aula, P., and W. W. Nichols (1967). J. Cell. Physiol., 70, 281.

El-Alfi, O. S., P. M. Smith and J. J. Biesele (1965). Hereditas, 52, 285.

Fogh, J., and H. Fogh (1965). Proc. Soc. Exptl. Biol. Med., 119, 233.

Harnden, D. G. (1964). Am. J. Human Genet., 16, 201.

Ignoffo, C. M. (1968). Current Topics in Microbiology, 42, 129.

Kihlman, B. A. (1966). Actions of Chemicals on Dividing Cells. Englewood Cliffs, N.J., Prentice Hall, Inc.

Kihlman, B. A., W. W. Nichols and A. Levan (1963). Hereditas, 50, 139.

Makino, S. and T. Aya (1968). Cytologia, 33, Nos. 3-4, 370.

Matsaniotis, N., F. Maounis, K. A. Kiossoglou, and D. E. Anagnostakis (1968). Lancet, May 4, 978.

Matsaniotis, N., K. A. Kiossoglou, F. Maounis, and D. E. Anagnostakis (1966). Lancet, Dec. 24, 1421.

Mella, B. and D. J. Lang (1967). Science, 155, 80.

Moorhead, P. S. (1970). Genetic Concepts and Neoplasia. Baltimore, The Williams and Wilkins Co.

Nichols, W. W. (1963). Hereditas, 50, 53.

Nichols, W. W. (1969). In Handbook of Molecular Cytology, A. Lima-De-Faria, editor. Amsterdam, London, North Holland Publishing Co.

Nichols, W. W. (1970). Ann. Review of Microbiology, 24, 479.

Nichols, W. W., A. Levan, W. Heneen, and M. Peluse (1965). Hereditas, 54, 213.

Paton, G. R., J. P. Jacobs, and F. T. Perkins (1965). Nature, 207, 43.

Sparrow, A. H. (1961). Mutation and Plant Breeding. Washington, D.C. National Academy of Sciences Publication #891.

Stich, H. F. and D. S. Yohn (1970). Progress in Medical Virology, 12, 78. J. L. Melnick, editor. Basel, New York. S. Karger.

Zur Hausen, H., and E. Lanz (1966). Z. med. Mikrobiol. u. Immunol., 152, 60.

# POSSIBLE RELATIONSHIPS BETWEEN MUTAGENESIS AND CARCINOGENESIS

*Isaac Berenblum*

The extensive knowledge that has accumulated during the past 40 years, from animal experimentation, about the nature and diversity of chemical carcinogens (Hartwell, 1951; Hueper and Conway, 1964; Arcos et al., 1968) owes its inspiration to previously acquired information about environmental cancer in men.

So long as these examples of human cancer were only thought of in terms of occupational hazards, the implications with respect to cancer prevention seemed of limited importance and remote from the generality of human cancers. But a radical change in outlook came about with the subsequent realization that extrinsic causes of human cancer could also be non-occupational in origin--in the form of contaminants in the atmosphere, artificial additives to food, certain domestic products (pesticides, etc.), the consequence of certain habits (cigarette smoking, etc.), and even including naturally-occurring substances produced by fungi (e.g., aflatoxin) affecting food crops and derivatives of wild plants (e.g., Senecio alkaloids) taken in the form of herbal medicines. Added to this is the more recent indication that some viral infections might also contribute to cancer development in man, this being still a matter of surmise, however, though based on well-established evidence in animals.

Thus, the earlier belief that less than five percent of all human cancers could be attributed to extrinsic causes has had to be drastically revised. According to more recent assessments (World Health Organization, 1964) at least 75 percent of all human cancers are likely to have an environmental component. The significance of this is that cancer prevention, instead of being thought of as a mere adjunct to cancer therapy, must now be considered a major goal in the fight against cancer in man.

The "Somatic Cell Mutation Theory of Cancer" (Boverie, 1929; Bauer, 1928), which seeks to explain carcinogenesis as resulting from a change in the genetic makeup of the cell, was actually postulated long before there was knowledge of the chemical nature of carcinogens and was originally introduced to account for two striking features of cancerous growth in man--the unlimited variety of tumor types and the fact that, on cell division, the daughter cells maintained their neoplastic properties.

For a time, it was thought theoretically impossible to prove or disprove this attractive theory of tumor causation, since the acid test of a mutation -- crossing the alleged mutated cell with a normal cell and then analyzing the progeny for Mendelian segregation -- could not be applied to somatic cells (Haldane, 1934). We now know that this argument is no longer valid.

Indirect methods of testing the theory were meanwhile attempted: (a) by seeking a correlation between mutagenic and carcinogenic action among compounds that were already known to have one or the other property, and (b) by analyzing cancer incidence curves with respect to age, to determine whether a mathematical formula could be found to fit the data, on the assumption that a sequence of random mutations were responsible for tumor development.

Regarding the first method of approach, some of the early tests did seem to show a correlation between mutagenic and carcinogenic activity for a number of compounds, though with some notable exceptions (Burdette, 1956). It should, however, be pointed out in advance that an absolute correlation between the two actions could hardly be expected in practice, even if the theory were correct, for the following reasons:

1. Mutagenic action covers a wide range of effects, from deletions of whole chromosomes to point mutations. It can be expressed in many different ways (morphologically, biochemically, serologically,

etc.), not all of which are demonstrable by available testing methods.

2. Agents may exhibit mutagenic effects by one test system, but not by another (e.g., in Drosophila, Neurospora, E. coli, phage, etc.). Mutagenic propensities may be missed unless tests are carried out on many different systems (Fishbein et al., 1970; Vogel and Rohrborn, 1970; Mrak, 1969).

3. Carcinogenic action may likewise operate in one test system (e.g., by systemic administration, resulting in tumor induction restricted to the liver, lungs, mammary tissue, etc.) and not in another test system (e.g., by skin application or subcutaneous injection) (Hartwell, 1951; Hueper and Conway, 1964; Arcos et al., 1968). The claim that a particular compound is non-carcinogenic may only mean that it has not been tested under sufficiently diverse conditions.

4. Mutagenic and carcinogenic activity may both depend on metabolic conversion of the compounds being tested. This may vary according to the species, strain, and sex of the test animal or tissue and can also be influenced by extrinsic factors.

The alternative approach for testing the validity of the "Somatic Cell Mutation Theory of Cancer" -- analyzing the cumulative cancer incidence curves with respect to age and devising a mathematical formula to fit the data--has given rise to the hypothesis that two consecutive random mutations, in the case of experimental carcinogenesis in mice (Charles and Luce-Clausen, 1942; Iversen and Arley, 1950), and five to six for cancer development in man (Fisher and Hollomon, 1951; Nordling, 1953; De Waard, 1964) are involved. But this only means that a theory postulating multiple random mutations for cancer development is consistent with the mathematical expression, it does not afford proof that the theory is correct. In fact, from our knowledge of the two-stage mechanism of skin carcinogenesis in mice, it would seem that the initiating phase might, in theory, be mutational (Berenblum and Shubik, 1949; Gelboin et al., 1965) though even this has not, so far, been substantiated experimentally (Trainin et al., 1964). The promoting phase during the long latent period of carcinogenesis, responsible for the "phenotypic" expression of the tumor, is almost certainly not mutational in character (Berenblum, 1954).

I have so far dealt with the problem from the viewpoint of the validity of the "Somatic Cell Mutation Theory of Cancer," as an attempt to explain the mode of origin of the disease. There is, however, another aspect to the problem which is of more immediate, practical importance and which is of particular interest to the present discussion, i.e., to decide (a) whether the correlation between mutagenesis and carcinogenesis is sufficiently close for the former to serve as an index of the latter with regard to testing procedures, and (b) whether it is possible to predict carcinogenic hazards in man on the basis of evidence of mutagenesis. The remainder of this discussion will concern itself more specifically with these practical objectives.

When the question of a possible correlation between mutagenesis and carcinogenesis was first seriously considered in the early fifties, several thousand compounds had already been tested for carcinogenic activity in animals, of which more than 300 had been found to be carcinogenic (Hartwell, 1951). Only a dozen or so of these compounds had been tested for mutagenic action (Burdette, 1956). It was perhaps naive to have concluded that some correlation existed between mutagenic and carcinogenic action. Since then, there has been a growing interest in the problem of mutagenesis, both from the viewpoint of fundamental genetic studies and from practical considerations of food additives, pesticides, etc. (Fishbein et al., 1970; Vogel and Rohrborn, 1970; Mrak, 1969). Tests for mutagenesis in certain experimental organisms can be performed more rapidly than tests for carcinogenesis, and an abundance of data on mutagenesis is now available.

The list of mutagenic agents (see Fishbein and Nichols, this volume) now includes even such simple compounds as hydrogen peroxide, ethylene oxide, formaldehyde, chloral hydrate, and inorganic salts--all of which are non-carcinogenic--as well as simple lactones, epoxides, sulphur- and nitrogen-mustard, urethane, etc.--which are carcinogenic. Of the most potent carcinogens, some (notably among the polycyclic aromatic hydrocarbons) have weak, if any mutagenic activity. The fact that the closest correlation between mutagenic and carcinogenic activity is found among the alkylating agents, which are capable of reacting with DNA, is an interesting theoretical point but hardly one that serves as an indication of a general interdependence between the two forms of action. Nor is it suprising that there is a closer correlation between mutagenic and tumor-inhibiting action than between mutagenic and carcinogenic action, seeing that most cancer chemotherapeutic agents were initially chosen for their ability to interfere with the genetic apparatus of the cell.

In short, far from finding a close correlation between mutagenesis and carcinogenesis, one is led to conclude that almost any chemically-reactive compound, if tested by a sufficiently wide range of methods, is likely to prove mutagenic. This does not apply to carcinogenesis.

As regards the mechanisms of action in the two cases and their relevance to possible harm for man, serious consideration must be given to the fact that both mutagenesis and carcinogenesis are essentially irreversible processes. One would, therefore, hesitate to ignore weak effects in either case or assume that there is a safe threshold dose for either. On the other hand, it would be irrational to suppose that the same element of risk (for man) applies to both. Carcinogenic action leads to tumor formation and is, therefore, a serious potential hazard, however weak the action. Mutagenesis, unless associated with carcinogenesis, is a potential source of danger mainly when affecting the germ cells in the testis and ovary, i.e., with the effect passing (usually as a recessive character) to the offspring. (This is, admittedly, an oversimplification of the case, since somatic mutations in man can, hypothetically, cause demonstrable changes in the body as a whole, when affecting progenitor stem cells--notably in the haematopoietic and lymphopoietic systems--or as dominant lethal mutations in ova, leading to loss of fertility.)

This raises an important tactical problem, in connection with the testing of mutagenic activity as an index of a potential hazard to man, i.e., the question whether equal weight should be given to evidence of mutagenic activity (a) in unicellular organisms, (b) in plants, (c) in tissue culture, (d) in _Drosophila_ or other lower multicellular organisms, (e) in somatic cells of higher animals, (f) in germ cells of higher animals, and (g) in man (Vogel and Rohrborn, 1970). I realize that to narrow down one's criteria for mutagenesis could lead to a false sense of security with respect to man. But one could equally well argue that, by accepting all evidence of mutagenic activity in whatever way it manifests itself, the situation might arise where every conceivable substance which is foreign to the body would have to be excluded from man's environment.

It may well be that I am overstressing the essential difference and lack of association between carcinogenesis and mutagenesis. From a strictly practical viewpoint, such an exaggerated stand might be justified. A strong case may, therefore, be made for adopting different criteria for carcinogenesis and mutagenesis, insofar as the data are to apply to potential human hazards. In the case of carcinogenesis, any positive findings, however weak and however restricted the action, must be interpreted as a potential source of danger. The main question that arises is whether the usefulness of the substance (e.g., for therapeutic purposes) outweighs the potential carcinogenic hazard. In the case of mutagenesis, evidence of risk to man himself must decide the issue.

If viewed in this light, the question of a correlation between carcinogenesis and mutagenesis becomes an exercise in theoretical speculation and of little relevance to the practical problem of human hazards.

## References


Arcos, J. C., M. F. Argus, and G. Wolf (1968). _Chemical Induction of Cancer_ (2nd Edition). New York, Academic Press.

Bauer, H. K. (1928). _Mutationstheorie der Geschwulst-Entstehung_. Berlin, Springer.

Berenblum, I. (1954). _Advanc. Cancer Research_, _2_, 129.

Berenblum, I. and P. Shubik (1949). Brit. J. Cancer, 3, 109.
Boverie, T. (1929). The Origin of Malignant Tumors. Baltimore, The Williams and Wilkins Co.
Burdette, W. J. (1956). Cancer Research, 15, 201.
Charles, D. R., and E. M. Luce-Clausen (1942). Cancer Research, 2, 261.
De Waard, R. H. (1964). Internat. J. Rad. Biol., 8, 381.
Fishbein, L., W. G. Flamm, and H. L. Falk (1970). Chemical Mutagens. New York, Academic Press.
Fisher, J. C., and J. H. Hollomon (1951). Cancer, 4, 916.
Gelboin, H. V., M. Klein, and R. R. Bates (1965). Proc. Nat. Acad. Sci., 53, 1353.
Haldane, J. B. S. (1934). J. Path. and Bact., 38, 507.
Hartwell, J. L. (1951). Survey of Compounds Which Have Been Tested for Carcinogenic Activity (2nd Edition). U. S. Public Health Service, Publ. No. 149. See also Shubik, P., and J. L. Hartwell, 1959 (Suppl. 1) and 1969 (Suppl. 2).
Hueper, W. C.., and W. D. Conway (1964). Chemical Carcinogenesis and Cancer. Springfield, Ill., Charles C. Thomas.
Iversen, S., and N. Arley (1950). Acta Path. and Microbiol. Scand., 27, 773.
Mrak, E. M. (1969). Report of the Secretary's Commission on Pesticides and Their Relationship to Environmental Health. Dept. of Health, Education, and Welfare, Washington, D.C., U. S. Government Printing Office.
Nordling, C. O. (1953). Brit. J. Cancer, 7, 68.
Trainin, N., A. M. Kaye and I. Berenblum (1964). Biochem. Pharmacol., 13, 263.
Vogel, F., and G. Rohrborn, editors (1970). Chemical Mutagenesis in Mammals and Man. Berlin, Springer-Verlag.
World Health Organization (1964). Cancer Prevention, Rep. Ser. No. 276.

# INTERRELATIONS BETWEEN CARCINOGENICITY, MUTAGENICITY, AND TERATOGENICITY

*James G. Wilson*

The three toxicologic processes named in the title share the doubtful distinction of being among the most potent of the "alarm words" in the rich vocabulary currently used to describe real and imagined risks from the environment. A journalist or commentator who mentions any one of them in connection with a newly discovered environmental change is assured of some attention from the confused public, and if he is clever enough to use all three in the same context, it may make his reputation. Of the three, teratogenicity is probably the most frightening to the average citizen because it conjures up imagery not only of suffering and death but, more pathetically, of deformed children, of lifelong invalids, and of whole families blighted by heavy financial and emotional burdens.

It is not my purpose, however, to dwell on the public relations aspect of the subject, but instead to examine the interrelations between these processes, particularly as they may concern environmental contaminants, with as much scientific objectivity as possible. Carcinogenicity, mutagenicity and teratogenicity are said to share features other than that mentioned above. These points of similarity were summarized in the Corvallis Task Force Report (1970) as: (1) insidiousness of nature, (2) relatively long time lag between exposure and overt effect, (3) irreversibility of the diseases, (4) relatively great susceptibility of immature or developing tissues, (5) some similarity of etiologic factors, and perhaps other less obvious similarities. In some respects, however, the ostensibly shared features are more apparent than real when scrutinized closely.

As to the first of the features, insidiousness, there can be no doubt but that in all three cases the causes seem disproportionately mild in view of the severity of the effect in many instances, but this is not an unusual phenomenon in toxicology. It is noteworthy, however, that some mutations are not detrimental, a fact that is basic to the process of evolution.

Regarding the duration of the time lag between exposure to the cause and recognition of an effect in these disease states, however, there may be orders of magnitude of difference. In teratogenesis the effect can be apparent within hours of exposure but is typically evident at or soon after birth, when the adverse influence occurred during the same gestation; thus elapsed time from induction to diagnosis is generally a few weeks or months and rarely more than a few years. The corresponding interval for carcinogenesis can range from a few months to many years. Mutagenesis may be manifested in the succeeding generation of progeny if dominant or sex-linked or involving chromosomal aberration, two or more generations later if a common recessive, or perhaps never if a rare recessive.

Irreversibility, unfortunately, is the prevailing nature of all three types of disease under discussion although, if medical treatment is accepted as "reversibility", then some prognostic differences exist. Many cancers are subject to surgical removal or chemotherapeutic reversal; many developmental errors can be corrected by surgery or, if of certain metabolic types, alleviated by drug or dietary therapy. But for mutations, the underlying defect persists and may become manifest whenever reproduction occurs, although excision of or substitution for defective genes may offer hope for the future (Lederberg, 1970).

Greater susceptibility of immature tissues is a feature that is inherent in one, shared to a degree by another, and possibly lacking in the third class of effects being discussed. None but immature tissues are susceptible to teratogenesis, and in general the more immature the target tissue, the greater the susceptibility. Certain cancers have been shown to be more easily induced in newborn or infant animals than in adults, but this is by no means universally true (Toth, 1968). Some types of neoplasms occur mainly in young individuals, but many others are more prevalent in mature or even senile individuals. The latter should be qualified by noting that cancers in older individuals most often occur in tissues that retain a generative function, e.g., bone marrow, basal layer of skin, and crypt cells of intestine. Gene mutations have not been shown to be more easily induced in immature individuals or germ cells than in mature ones and, on the contrary, one study revealed that dominant lethals in mice were more often induced by alkane sulfonic esters in mature sperm than in spermatids (Ehling et al., 1968).

Similarity of etiologic factors, among the presumed common features shared by carcinogenesis, mutagenesis and teratogenesis, can be substantiated only to a limited extent. An indisputable example is radiation, which over the years has become a classic means for the laboratory induction of all three types of lesions. Not only was it one of the first causative agents to be identified in each of the fields, it remains as a useful experimental tool because of the precision with which it can be applied. In addition, a few chemical agents such as urethan, nitrogen mustard, aflatoxin and benzo(a)-pyrene are now known to be active in mammals in all three areas, but beyond this the case for common causation is not easily supported. Dulbecco (1964) in summarizing a Biology Research Conference stated, "Most substances which are highly mutagenic are not carcinogenic and vice versa. For instance, among the acridines some are mutagenic, others carcinogenic, but none are both; among the alkylating agents we find the same situation, with the exception of two classes. The most powerful carcinogens, such as aromatic hydrocarbons, are not mutagenic, or at most, very weakly so." Although no longer strictly accurate in detail, the sense of this statement is applicable today in comparing carcinogenic or mutagenic chemicals with teratogenic chemicals. In fact, the acridines mentioned by Dulbecco as being either mutagenic or carcinogenic have on several occasions been reported to be nonteratogenic, whereas the alkylating agents, also found to be ambiguous regarding mutagenesis and carcinogenesis, have always proven to be teratogenic when appropriately tested. Actually, very few carcinogenic chemicals have been shown to be teratogenic but, paradoxically, the vast majority of anti-cancer drugs are teratogenic (Chaube and Murphy, 1968; Tuchmann-Duplessis, 1969). It has come to be axiomatic in teratology that such drugs as antibiotics, antimalarials, anthelmintics, fungicides and insecticides, as well as the diverse groups of antineoplastic drugs, are suspect as being teratogenic until proven otherwise, because by design these compounds interfere with basic metabolic processes in rapidly proliferating systems. There are few tissues, including neoplasms, that approach the rapid proliferation shown by the embryo during early organogenesis.

To further document the fact that commonness of causation is not as prevalent as has been suggested and to avoid the tedium of agent-by-agent comparison, I should like to summarize an exhaustive analysis by Harold Kalter (1971). He has reviewed 86 compounds thought to be either mutagenic or teratogenic in mammals. Some evidence was found for mutagenicity of 36 and teratogenicity of 40, but only 15 of the 86 have been shown to have both properties. Only three of the 15, namely, aflatoxin, benzo(a)pyrene and nitrogen mustard have also been shown to be carcinogenic, although it must be noted that all have not been rigorously tested.

The case for common causation is no more convincing as regards viruses. Approximately a half dozen viruses are established as teratogenic in mammals, of which two (rubella and cytomegalovirus) are effective in man. It is increasingly clear that a broad range of viral agents of both RNA and DNA varieties give rise to neoplasms in animals, but to date no firm evidence has shown a viral agent to be responsible for a malignant tumor in man, although there is reason to believe that Burkitt's tumor, a lymphoma in African children, is a likely candidate (Report of Panel on Carcinogenesis, 1970). Viruses have not been established as a causative factor in gene mutations, although they may contribute to chromosomal abnormality and thus through this mechanism, e.g., translocations, cause a small proportion of heritable disease (Nichols, 1966).

Other proposed points of similarity among carcinogenesis, mutagenesis, and teratogenesis are more tenuous than those mentioned above and would not add appreciably to the present discussion. Before leaving the subject, however, it should be mentioned that a low level of epidemiological concurrence has been observed between certain childhood cancers and congenital defects in man (Miller, 1968), but the significance of this remains obscure. Therefore, on the basis of present evidence, the features shared by the three disease states are at a somewhat superficial level and, if underlying commonness of mechanisms exists, it is yet to be revealed.

Now to consider the obverse point of view, how are carcinogenesis, mutagenesis and teratogenesis different? One way to answer this question is to describe the three conditions in basic and unqualified terms (Figure 1). Accordingly, carcinogenesis results when proliferating cells are released from the restraints imposed on parent or host tissues; mutagenesis begins with a transmissable change in the nucleotide sequence or number of one or more strands of chromosomal DNA; and teratogenesis is deviant development of such magnitude as to change significantly the final structural or functional makeup of the individual. Although these simple statements are inadquate as definitions in that they fail to allow for exceptions and special situations, they serve the useful purpose of showing that the three conditions seem to be concerned with different processes at different operational levels. To further emphasize known differences, it can be said that one is uncontrolled proliferation at the cellular level, one is altered nucleotide sequence or amount at the molecular level, and one is change in developmental pattern at the tissue or organ level; but it is recognized that all may ultimately be explicable in molecular terms.

Because the final manifestations of carcinogenesis, mutagenesis and teratogenesis may be quite divergent, and since there is in fact little similarity in causation, it seems logical to examine whatever degree of specificity may exist in each category between causative agents and type of response, e.g., cigarette smoking is primarily associated with respiratory cancer. Although the subject has not been widely studied, available data

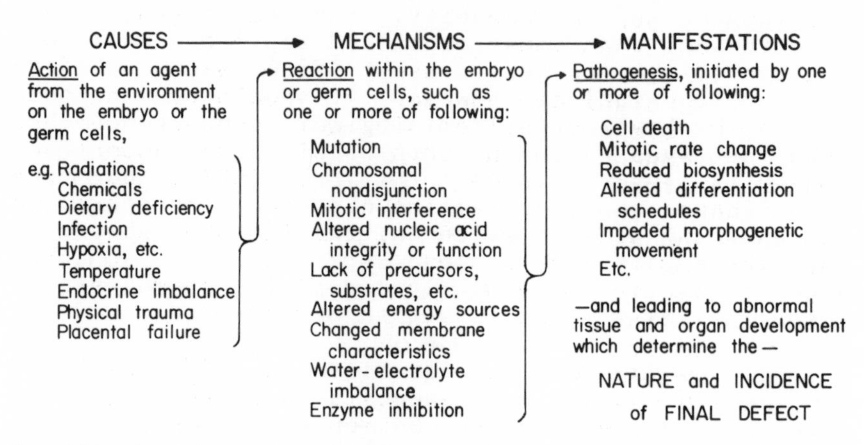


Fig. 1. This scheme depicts the complex sequence of events in teratogenesis. Although some parts may be held in common with carcinogenesis (e.g., certain causes) and mutagenesis (e.g., somatic mutations and chromosomal aberrations), many aspects are unique to teratogenesis, thus further emphasizing the basic differences existing among these process. Reprinted with permission from Wilson, 1971.

suggest that mutagenic agents do not often show agent-response specificity. In other words, mutations tend to affect the genome randomly, although preferential sites for mutations are well known in lower organisms (Jukes, 1966). In carcinogenesis the question is complicated by the fact that some carcinogens affect different tissues in different ways depending on the conditions under which the agent is applied, but many and possibly most affect predictable target tissues by inducing specific types of neoplasms. Examples of agent-response specificity, in addition to that cited above, are: (1) lymphatic and skin cancer after irradiation, (2) mammary carcinoma in mice associated with viral infections, (3) urinary bladder cancer with aromatic amines, (4) lung cancer with airborne asbestos, and (5) skin cancer with polycyclic aromatic hydrocarbons. In teratogenesis a high degree of specificity between the nature of the agent and the type of malformation produced has been frequently demonstrated (Wilson, 1961). An even greater degree of specificity in teratology seems to exist between the developmental stage when the agent is applied and the type of abnormality produced, a situation with no close parallel in either mutagenesis or carcinogenesis.

The most important area of difference between the three disease processes may be found in their basic toxicologic characteristics. It has already been mentioned, but needs particular emphasis here, that teratogenesis is an acute toxicologic phenomenon, requiring only short or instantaneous exposure and being manifested within a relatively short time thereafter. Mutations also may be instantaneously induced, as with X-rays, but in other situations they seem to be the consequence of continuous or repeated exposure, and it is not clear whether the effect is a matter of cumulative dosage or one of increasing probability of a "hit." Of course the final expression of mutagenesis can be delayed in the extreme. Carcinogenesis is usually regarded as one of the standard forms of chronic toxicity.

Zero tolerance is a controversial concept which implies that an agent found toxic at any dose is unsafe at all doses, regardless of how small, and consequently lacks a no-effect range of dosage. The original evidence supporting this view came from the field of radiation-induced mutations. Authorities such as Herman J. Muller have for many years contended that there is no safe dose of ionizing radiation as far as mutation is concerned. Without definitive proof of the validity of this notion, it is often assumed to apply to mutagenic agents generally. It has also been extended to the area of carcinogenesis, and in 1958 it achieved legal recognition in the form of the Delaney amendment, which in effect denied the existence of a no-effect dose for the carcinogenicity of food additives. There may be justification for this action as it applies to unessential chemicals such as certain food additives, cosmetics, and environmental pollutants, but this attitude should not be allowed to override honest attempts to balance known benefit against possible risk, a consideration involved in the use of many drugs and pesticides.

Recent suggestions that the concept of zero tolerance be extended to include teratogenicity (Epstein, 1970) must be challenged on scientific grounds. Not only is there lack of evidence to support such a suggestion but, on the contrary, there are voluminous experimental data showing that teratogenic agents can, under well controlled conditions, usually be shown to have a no-effect range of dosage. Although the existence of no-effect levels in teratogenesis has not been the subject of extensive literature review, it has been discussed in several connections (Wilson, 1964, 1968) and was accepted as a basic principle in teratology by various groups that have considered the problem of teratological testing, such as a World Health Organization Scientific Group (1966) and the Panel on Reproduction of the Food and Drug Administration's Advisory Committee on Protocols for Safety Evaluation (1969).

The concept of a zero tolerance, regardless of its applicability to carcinogenesis and mutagenesis, is subject to yet another logical objection in its application to teratogenesis. In contemplating the remarkable variety of chemicals that have been demonstrated to be teratogenic, albeit sometimes at extremely high doses, teratologists have quite naturally raised the question, "Are not all chemicals teratogenic if administered in sufficient dosage at the most sensitive time in development to the most susceptible species?" This question cannot be answered definitively because of the technical impossibility of performing all of the requisite tests, but it is given theoretical credence by experienced investigators in the field, despite the fact that many agents that might be teratogenic cannot be shown to be so because appropriate conditions of dosage would preclude or terminate pregnancy. Thus the statement that all chemicals are potentially teratogenic is an unproven generalization, but the fact remains that a larger number and variety of agents have been shown to be teratogenic than either carcinogenic or mutagenic. In any case, because so great a number of useful chemicals, including many drugs, are either known to be or are reasonably suspect of being teratogenic at high dosage, any extension of the zero tolerance concept as exemplified by the Delaney amendment to include teratogenic agents would be both scientifically indefensible and therapeutically unfortunate.

In various connections the question has been raised as to the feasibility of combining into a single test the protocols for safety evaluation of chemicals for carcinogenesis, mutagenesis, and teratogenesis. Even if entirely satisfactory tests were now available in each of these separate areas, which of course is not the case, it would be impossible to meld them into a single test without compromising specific procedures now used to enhance sensitivity for the separate types of pathology. The conditions of exposure often determine whether, or the extent to which, a given system responds to one or another of the three types of agents, as summarized in Table I. For example, the most sensitive test in each case would presumably be one involving direct application of the test substance to the most susceptible tissue of a test animal. With carcinogenesis, this is often both practical and preferable, but with mutagenesis and teratogenesis, tests involving chemical agents are most meaningful when the agent is absorbed, distributed and metabolized by the test animal in a manner simulating that anticipated in human exposure. Furthermore, a test for teratogenesis would only be applicable to pregnant mammals and then would only have maximum predictive value when given for a short time during early organogenesis of the embryo.

Another illustration of the need to consider conditions of exposure involves duration of treatment (Table I). A realistic test for carcinogenesis or mutagenesis could be carried out by repeated treatments over a period of weeks or months with a

Table 1. Conditions of Exposure That Determine the Response in Carcinogenesis, Mutagenesis, and Teratogenesis

| | Susceptible Tissues | Optimal Time of Exposure | Duration and Level of Dosage |
|---|---|---|---|
| Carcinogen | Proliferating tissues | Uncertain, probably all stages capable of mitosis | Usually chronic, possibly all doses |
| Mutagen | Germinal tissues | All stages of gametogenesis | Either acute or chronic, possibly all doses |
| Teratogen | Possibly all immature tissues | Highest during early differentiation | Acute only, above usual no effect level |

dose that causes no other toxicity. In testing for teratogenesis, however, two or more doses with the highest at a known maternal effect level should be used, and these should be administered for only a few days corresponding with the most sensitive period of the embryo. This time limitation is necessary to avoid the possibility of either inducing or inhibiting catabolic enzymes or other adaptative systems now known to be capable of altering maternal blood levels after repeated dosage (King et al., 1965). Chronic treatment begun prior to or at conception could result in an appreciably different dose response by the embryo than if treatment were begun when sensitivity is greatest during early embryogenesis, e.g., days 9 to 10 in rat, days 20 to 25 in rhesus monkey. Examination of Table I suggests additional reasons but the foregoing should be sufficient to establish the point that carcinogenesis, mutagenesis and teratogenesis cannot be evaluated simultaneously without sacrifice of reliability in one or all of the areas concerned.

## Conclusion

An attempt to identify common features in carcinogenesis, mutagenesis, and teratogenesis leads to the conclusion that the similarities are at a superficial level and that these processes are basically different. For both scientific and logical reasons, the concept of zero tolerance cannot be extended to include teratogenesis. It is emphasized that protocols for testing of carcinogenesis, mutagenesis, and teratogenesis cannot be combined into a single test without compromising further the limited sensitivity of existing test methods.

## References

Chaube, S. and M. L. Murphy (1968). In Advances in Teratology, Vol. 3, D. H. M. Woolam, editor. New York, Academic Press.

Dulbecco, R. (1964). J. Cell and Comp. Physiol., 64 (Suppl. 1), 181.

Ehling, U. H., R. B. Cumming, and H. V. Malling (1968). Mutation Res., 5, 417.

Epstein, S. (1970). Nature, 228, 816.

Jukes, T. H. (1966). Molecules and Evolution. New York, Columbia University Press.

Kalter, H. (1971). In Chemical Mutagens, A. Hollaender, editor. New York, Plenum Press.

King, C. T. G., S. A. Weaver, and S. A. Narrod (1965). J. Pharm. and Exp. Ther., 147, 391.

Lederberg, J. (1970). BioScience, 20, 1307.

Miller, R. W. (1968). J. Natl. Cancer Inst., 40, 1079.

Nichols, W. W. (1966). Am. J. Human Genet., 18, 81.

Report of the Corvallis Task Force (1970). In Man's Health and the Environment--Some Research Needs. Washington, D. C., U. S. Government Printing Office.

Report of the Panel on Carcinogenesis (1970). P. Shubik, Chairman. Advisory Committee on Protocols for Safety Evaluation, Food and Drug Administration, Washington, D. C.

Report of the Panel on Reproduction (1969). Advisory Committee on Protocols for Safety Evaluation, Food and Drug Administration, Washington, D.C.

Toth, B. (1968). Cancer Res., 28, 727.

Tuchmann-Duplessis, H. (1969). In Teratology, Proceedings of the International Symposium on Teratology at Como, Italy, A. Bertelli, editor. Amsterdam, Excerpta Medica Foundation.

Wilson, J. G. (1961). In Congenital Malformations. Philadelphia, J. B. Lippencott.

Wilson, J. G. (1964). J. Pharmacol. and Exp. Therap., 144, 429.

Wilson, J. G. (1968). In Toxic Effects of Anesthetics, B. R. Fink, editor. Baltimore, Williams and Wilkins Co.

Wilson, J. G. (1972). In Pathophysiology of Gestation, Vol. 2, N. S. Assali, editor. New York, Academic Press (in press).

World Health Organization (1967). Technical Report No. 346. Geneva, W.H.O.